Environmental Health Criteria 17

MANGANESE

Published under the joint sponsorship of
the United Nations Environment Programme,
the International Labour Organisation,
and the World Health Organization

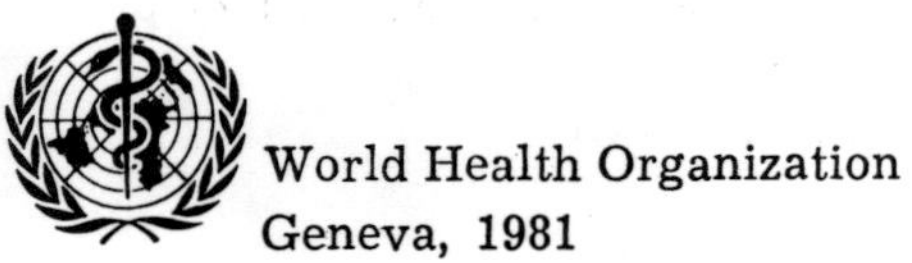

World Health Organization
Geneva, 1981

ISBN 92 4 154077 X

© World Health Organization 1981

PRINTED IN FINLAND

— VAMMALAN KIRJAPAINO OY, VAMMALA —

CONTENTS

NOTE TO READERS OF THE CRITERIA DOCUMENTS

While every effort has been made to present information in the criteria documents as accurately as possible without unduly delaying their publication, mistakes might have occurred and are likely to occur in the future. In the interest of all users of the environmental health criteria documents, readers are kindly requested to communicate any errors found to the Division of Environmental Health, World Health Organization, Geneva, Switzerland, in order that they may be included in corrigenda which will appear in subsequent volumes.

In addition, experts in any particular field dealt with in the criteria documents are kindly requested to make available to the WHO Secretariat any important published information that may have inadvertently been omitted and which may change the evaluation of health risks from exposure to the environmental agent under examination, so that the information may be considered in the event of updating and re-evaluation of the conclusions contained in the criteria documents.

WHO TASK GROUP ON ENVIRONMENTAL HEALTH CRITERIA FOR MANGANESE

Members

Dr M. Cikrt, Institute of Hygiene and Epidemiology, Prague, Czechoslovakia
Dr G J. van Esch, Toxicology and Food Chemistry, National Institute of Public Health, Bilthoven, Netherlands (*Chairman*)
Dr G. F. Hueter, Environmental Research Center, US Environmental Protection Agency, Research Triangle Park, NC, USA
Dr I. C. Munro, Toxicology Research Division, Bureau of Chemical Safety, Department of National Health and Welfare, Ottawa, Ontario, Canada (*Rapporteur*)
Dr H. Oyanguren, Institute of Occupational Health and Air Pollution, National Health Service, Santiago, Chile
Dr M. Šarić, Institute of Medical Research and Occupational Health, Zagreb, Yugoslavia
Dr S. Šigan, Sysin Institute of General and Community Hygiene, Moscow, USSR
Dr N. Skvortsova, Laboratory for Air Pollution Control, Sysin Institute of General and Community Hygiene, Moscow, USSR
Professor M. Tati, Department of Public Health, Gifu University Medical School, Gifu, Japan
Dr I. Ulanova, Institute of Industrial Hygiene and Occupational Diseases, Moscow, USSR (*Vice-Chairman*)

Representatives of other agencies

Dr H. M. Mollenhauer, Division of Geophysics, Global Pollution and Health, United Nations Environment Programme, Nairobi, Kenya
Dr D. Djordjević, Occupational Health and Safety Branch, International Labour Organisation
Mrs M. Th. van der Venne, Health Protection Directorate, Commission of the European Communities, Luxembourg

Secretariat

Dr Y. Hasegawa, Medical Officer, Control of Environmental Pollution and Hazards, World Health Organization, Geneva, Switzerland
Dr H. de Koning, Scientist, Control of Environmental Pollution and Hazards, World Health Organization, Geneva, Switzerland
Dr J. E. Korneev, Scientist, Control of Environmental Pollution and Hazards, World Health Organization, Geneva, Switzerland
Dr G. E. Lambert, Scientist, Occupational Health, World Health Organization, Geneva, Switzerland
Dr B. Marschall, Medical Officer, Occupational Health, World Health Organization, Geneva, Switzerland
Dr V. B. Vouk, Chief, Control of Environmental Pollution and Hazards, World Health Organization, Geneva, Switzerland (*Secretary*)

A WHO Task Group on Environmental Health Criteria for Manganese met in Geneva from 22 to 26 September 1975. Dr B. H. Dieterich, Director, Division of Environmental Health, opened the meeting on behalf of the Director-General. The Task Group reviewed and revised the second draft of the criteria document and made an evaluation of the health risks from exposure to manganese and its compounds.

The first and second drafts of the criteria document were prepared by Dr P. S. Elias of the Department of Health and Social Security, London, England. The first draft was based on national reviews received from the national focal points for the WHO Environmental Health Criteria Programme in Bulgaria, Japan, New Zealand, the United Kingdom, the USA, and the USSR. The second draft was prepared according to comments received from national focal points in Canada, Chile, Czechoslovakia, Greece, Japan, Netherlands, New Zealand, Poland, Sweden, the USA, and the USSR; and from the Commission of the European Communities, the Food and Agriculture Organization of the United Nations, the Ethyl Corporation, the International Union of Biological Sciences, the International Union of Pure and Applied Chemistry, the United Nations Economic Commission for Europe, and the World Meteorological Organization. Dr P. S. Elias and Dr I. C. Munro, Bureau of Chemical Safety, Department of National Health and Welfare, Ontario, Canada, assisted the Secretariat in the preparation of a third draft, which was distributed for comments to the Task Group members. Additional comments on this draft were received from Dr R. J. M. Horton, US Environmental Protection Agency, Research Triangle Park, USA, and Professor M. Piscator, the Karolinska Institute, Stockholm, Sweden. Following the recommendations made by a WHO Consultative Group on the application of environmental health criteria, Bilthoven, Netherlands, 2—5 May 1977, a final draft was prepared by Dr H. Nordman, Institute of Occupational Health, Helsinki, Finland, taking into consideration the comments of members of the Task Group and of Professor P. S. Papavasiliou, the New York Hospital Centre-Cornell Medical Center, New York, USA, and Professor M. Piscator.

The collaboration of these institutions, organizations, and individual experts is gratefully acknowledged. The Secretariat wishes to thank, in particular, Dr P. S. Elias, Dr. I. C. Munro, and Dr H. Nordman for their help in the various phases of preparation of the document.

This document is based on original publications listed in the reference section but much valuable information was also obtained from publications reviewing and evaluating the essentiality and toxicity of manganese, including those by Cotzias (1958, 1962), Stokinger (1962), Schroeder et al. (1966), Suzuki et al. (1973a, 1973b, 1973c), WHO (1973), WHO Working Group (1973), US Environmental Protection Agency (1975), International Agency for Cancer Research (1976), and Šarić (1978). Owing to unforseen circumstances, it has not been possible to update the document beyond 1978.

Details of the WHO Environmental Health Criteria Programme, including some terms frequently used in the documents, can be found in the general introduction to the Environmental Health Criteria Programme published together with the environmental health criteria document on mercury (Environmental Health Criteria 1, Mercury, Geneva, World Health Organization, 1976) and now available as a reprint.

*   *
*

Financial support for the publication of this criteria document was kindly provided by the Department of Health and Human Services through a contract from the National Institute of Environmental Health Sciences, Research Triangle Park, North Carolina, USA — a WHO Collaborating Centre for Environmental Health Sciences.

1. SUMMARY AND RECOMMENDATIONS FOR FURTHER STUDIES

1.1 Summary

1.1.1 Analytical methods

Numerous analytical methods are available for the quantitative determination of manganese in environmental media and biological samples. The method the most frequently used is atomic absorption spectroscopy, which appears to be sufficiently sensitive for most analytical purposes. The way in which biological and environmental samples are procured and stored, prior to analysis, has an important bearing on the accuracy and validity of the results. For example, in air sampling, it is important to ensure that respirable particulate matter is collected. In the collection of biological samples with a low manganese content, contamination may constitute a major difficulty.

1.1.2 Sources and pathways of exposure

Manganese is one of the more abundant elements in the earth's crust and is widely distributed in soils, sediments, rocks, water, and biological materials. The major sources of man-made environmental pollution by manganese arise in the manufacture of alloys, steel, and iron products. Other sources include mining operations, the production and use of fertilizers and fungicides, and the production of synthetic manganese oxide and dry-cell batteries. Organomanganese fuel additives, though only a minor source at present, could significantly increase exposure, if they come into widespread use. Average manganese concentrations[a] in soils range from about 500 to 900 mg/kg and concentrations in sea water range from 0.1 to 5 μg/litre. Surface waters may have a manganese content of 1—500 μg/litre, but in areas where high concentrations of manganese occur naturally, levels may be considerably higher. Average manganese levels in drinking water range from 5 to 25 μg/litre.

Manganese is present in all foodstuffs, usually at concentrations below 5 mg/kg. However, concentrations in certain cereals, nuts, and shellfish can be much higher, exceeding 30 mg/kg in some cases. Levels in finished tea leaves may amount to several hundred mg/kg.

Manganese has been found in measurable quantities in practically all air samples of suspended particulate matter. Annual average levels in ambient air in unpolluted urban and rural areas vary from 0.01 to 0.07 μg/m³. However, in areas associated with the manganese industry, annual averages may be higher than 0.5 μg/m³, and have occasionally exceeded 8 μg/m³. About 80% of the manganese in suspended particulate matter is associated with particles having a mass median equivalent diameter (MMED)[b] of less than 5 μm, i.e., particles within the respirable range. This association with small particles favours the widespread airborne distribution of manganese.

1.1.3 Essentiality of manganese

Manganese is an essential trace element for both animals and man. It is necessary for the formation of connective tissue and bone, and for growth, carbohydrate and lipid metabolism, the embryonic development of the inner ear, and reproductive functions. Some specific biochemical functions of manganese have been discovered such as the catalysing of the glucosamine-serine linkages in the synthesis of the mucopolysaccharides of cartilage.

Estimates from intake and balance studies in man show that the daily requirement for adults is 2—3 mg/day and that of pre-adolescent children, at least 1.25 mg/day. Manganese deficiency states, which have been detected in a wide variety of animals, have been described only once in man, in association with vitamin K deficiency and the accidental omission of manganese from the diet. A distinctly negative manganese balance is found in newborn infants, the metal being excreted from stores that have accumulated in the

---

[a] Throughout the document, the term concentration refers to mass concentration, unless otherwise stated.
[b] Mass median equivalent diameter: equivalent diameter above and below which the weights of all larger and smaller particles are equal.

tissues during fetal life. However, deficiency symptoms have not
been detected.

1.1.4 Magnitude of environmental exposure

Food is the major source of manganese for man. Daily intake
ranges from 2 to 9 mg, depending on the relative consumption of
foods with a high manganese content, especially cereals and tea. In
young children and up to the age of adolescence, the daily intake
is about 0.06—0.08 mg/kg body weight; for breastfed and bottlefed
infants, it is only about 0.002—0.004 mg/kg body weight. Daily
intake with drinking water may range from a few micrograms to
200 μg, the average intake being about 10—50 μg/day.

The daily intake of manganese in the air by the general population in areas without manganese emitting industries is below
2 μg/day. In areas with major foundry facilities, intake may rise to
4—6 μg/day and in areas associated with ferro- or silicomanganese
industries it may be as high as 10 μg, with 24-h peak values exceeding 200 μg/day.

1.1.5 Metabolism

The respiratory and gastrointestinal tracts constitute the major
routes of absorption of manganese. Quantitative data are not available, but it seems unlikely that the skin is an important route of
absorption for inorganic manganese compounds, although organomanganese compounds can be absorbed by this route.

The extent of absorption of manganese following inhalation is
unknown. A certain proportion of inhaled manganese particles is
cleared by mucociliary action and swallowed, and is available for
gastrointestinal absorption. The small amount of information available concerning the gastrointestinal absorption of manganese in man
indicates that the absorption rate in healthy adults is below 5% but
that it is higher in anaemic subjects. This is supported by data from
studies on mice and rats. There is little information on gastrointestinal absorption in infants and children and not much is known
about the mechanism of absorption from the gastrointestinal tract.

In studies on experimental animals, preloading with high dietary
levels of manganese caused a decrease in the rate of absorption and
young rats appeared to have a considerably higher absorption rate
than adult rats.

The total manganese body burden for a man of 70 kg is about
10—20 mg. It is transported in the plasma bound to a β_1-globulin,
most likely transferrin, and is widely distributed throughout the
body. Manganese concentrates in tissues rich in mitochondria, the
highest concentrations being found in the liver, pancreas, kidney,
and the intestines. It can also penetrate both the blood-brain barrier

and the placenta. The disappearance half-time for manganese from the whole body is about 37 days and the half-time in the brain appears to be longer than that for the whole body. Tissue concentrations in man are remarkably stable throughout life. Variable excretion is known to play an important role in the homeostasis of manganese, but recent studies have shown that the variability of absorption is also important.

Inorganic manganese is mainly eliminated in the faeces. The principal route of excretion is with the bile, part of which is reabsorbed in the enterohepatic circulation. To some extent, manganese is also excreted with the pancreatic juice and through the intestinal wall; the importance of these routes may increase under abnormal conditions such as biliary obstruction or increased manganese exposure. It has been shown that only about 0.1—1.3% of the daily intake of inorganic manganese is normally excreted in the urine. However, larger amounts are excreted through the kidney following exposure to organomanganese tricarbonyl compounds, indicating that these compounds, which are used as additives in gasoline, are metabolized in the body.

1.1.6 Effects on experimental animals

The toxic effects of manganese on the central nervous system have been induced in various animal species, including the rat and monkey, mainly by the administration of manganese dioxide or dichloride. Exposure of a monkey to manganese dioxide aerosol, by inhalation, at concentrations of 0.6—3.0 mg/m³, for 95 1-h periods over 4 months, induced typical signs of central nervous system effects. Parenteral administration of manganese dioxide or dichloride also induced signs of central nervous system disturbance but oral administration produced fewer effects, presumably because of poor gastrointestinal absorption. Histopathological lesions found in intoxicated animals included degenerative changes, primarily in the striatum and pallidum, but lesions in the subthalamic nucleus, cortex, cerebrum, cerebellum, and the brain stem have also been observed. It has been shown that manganese causes depletion of dopamine, and probably serotonin, in the basal ganglia of monkeys, rabbits, and rats. These biochemical findings may explain, at least in part, the neurotoxic effects of manganese.

Inflammatory changes were produced in rats by intratracheal administration of manganese dioxide at concentration of 0.3 mg/m³ for 5—6 h daily, over 4 months; mottling was seen on the pulmonary radiographs of monkeys exposed to the same compound by inhalation (0.7 mg/m³). Sulfur dioxide was found to act synergistically with manganese dioxide on the respiratory tract of guineapigs.

Biochemical and histopathological changes have been reported in other organ systems, notably the liver. Testicular changes have been demonstrated in the rat after intravenous administration of permanganate at 50 mg/kg body weight and in the rabbit after administration of manganese dichloride at 3.5 mg/kg. Intraperitoneal injections of manganese(II) sulfate (10 mg/kg body weight, 15 injections) in mice increased the incidence of lung tumours; however, the carcinogenic, mutagenic, and teratogenic potential of manganese needs further investigation.

1.1.7   Effects on man

1.1.7.1 *Occupational exposure*

Chronic manganese poisoning is a hazard in the mining and processing of manganese ores, in the manganese alloy and dry-cell battery industries, and in welding. The disorder is characterized by psychological and neurological manifestations, the neurological signs closely resembling those that occur in other extrapyramidal disorders, notably parkinsonism. Autopsy reports on cases of chronic manganese poisoning have shown that lesions of the central nervous system are most severe in the striatum and pallidum, and may also be found in the substantia nigra. In one case, post-mortem analysis revealed a reduced concentration of dopamine. This finding combined with animal data and the fact that a precursor of dopamine, 3-hydroxy L-tyrosine (L-dopa), has been effective in the treatment of chronic manganese poisoning implicates the dopaminergic pathway in the etiology of extrapyramidal manifestations of the disease.

Individual susceptibility to the adverse effects of manganese varies considerably. The minimum dose that produces effects in the central nervous system is not known, but signs of adverse effects may occur at manganese concentrations in air ranging from 2 to 5 mg/m^3.

Although an increased incidence of pneumonia has repeatedly been reported in manganese workers, it is not possible to establish any exposure-effect relationships from available data. It may be that particle size distribution and the type of manganese compound are more important than the mass concentration of manganese in air. This may also be true for the nonspecific effects on the respiratory tract reported in manganese workers. Smoking appears to act synergistically with manganese in causing such effects.

The early diagnosis of manganese poisoning is difficult in the absence of reliable biological indicators of exposure. Repeated screening for subjective symptoms and thorough clinical examinations should be undertaken at regular intervals together with measurements of manganese in blood and urine. Measurement of manganese levels in faeces may serve as a useful guide to exposure.

With better understanding of the pathophysiology of manganese poisoning, new drugs have been introduced for its treatment. In many cases, the use of the dopamine precursor L-dopa, has been successful. The use of chelating agents has also been reported to have a beneficial effect, although sometimes only temporarily and mainly in the early stages of poisoning. This treatment cannot be expected to bring about any improvement in cases where structural neurological injury has already occurred.

1.1.7.2 *Community exposure*

Adverse effects have been reported in populations, in areas associated with manganese-processing plants. In 1939, increased morbidity and mortality due to lobar pneumonia were reported from Sauda in Norway, where a ferro- and silicomanganese plant was operating. The mortality rate was positively correlated with the amount of manganese alloy produced. Manganese was reported to occur in the ambient air as Mn (II, III) oxide (Mn_3O_4) at manganese concentrations of up to 45 $\mu g/m^3$. In another study, a higher prevalence of nose and throat symptoms and lowered respiratory function were registered in schoolchildren exposed to manganese concentrations in air ranging from 4 to 7 $\mu g/m^3$ (5-day mean values) compared with an unexposed control group. However, short-time sampling (1-h) of the factory smoke, down-wind, yielded a maximum level of 260 $\mu g/m^3$.

A 4-year study performed in a population living in the vicinity of a ferromanganese plant indicated that even a manganese exposure of only 1 $\mu g/m^3$ might be connected with an increase in the rate of acute respiratory disease. However, it is possible that some other factors, which were not sufficiently controlled, might have influenced the results.

In one study, the incidence of abortions and stillbirths was reported to be higher in wives of workers exposed to manganese for 10—20 years than in a control group. The study is difficult to evaluate as factors such as the occupations of the wives were not reported.

1.1.8 Organomanganese compounds

There are two groups of organomanganese compounds of toxicological importance. Manganese ethylene-bis-dithiocarbamate (Maneb) is used as a fungicide on edible crops. Toxicologically, the manganese fraction is of little importance, whereas the organic portion is part of a larger problem concerning this type of fungicide. The manganese tricarbonyl compounds constitute the other group of organomanganese compounds of toxicological significance. These are used as additives in unleaded petrol (gasoline) and future widespread use seems likely. After combustion, only a small fraction of the compound is emitted and this undergoes rapid photodecomposition to form compounds that, so far, have not been satisfactorily identified. Exposure to manganese tricarbonyl compounds is therefore likely to constitute an occupational hazard but community exposure to the parent compound will remain very small, even if the use of these compounds increases. Nevertheless, widespread use would result in increased community exposure to inorganic manganese and to other possible combustion products. Rats, hamsters, and monkeys have been exposed experimentally to combusted methylcyclopentadienyl manganese tricarbonyl (MMT) at concentrations of manganese in air ranging from 12 to 5000 μg/m³ for various periods ranging up to 66 weeks without any adverse effects. However, tissue levels of manganese increased in monkeys exposed to a manganese concentration in air of 100 μg/m³.

1.2 Recommendations for Further Studies

1.2.1 Analytical methods

There is a need for interlaboratory comparison to determine the accuracy of methods available for the estimation of manganese. Additional studies are required to determine particle size in airborne manganese particulate matter, so that total intake through the respiratory pathway can be estimated more precisely.

1.2.2 Environmental exposure

More precise data are needed on manganese intake, especially by inhalation. A better understanding of the translocation of manganese in the environment and factors that affect this process is required and its potential for bioaccumulation in environmental compartments should be explored in more depth.

1.2.3 Metabolism

Chemobiokinetic studies are necessary to identify, more precisely, the mechanisms involved in the uptake and clearance of manganese from the gastrointestinal tract and the respiratory system in both experimental animals and exposed populations and to obtain a better understanding of factors that affect these processes. Tissue levels at which adverse effects are observed should be established and special attention should be paid to the role of nutritional status and age in the metabolism of manganese.

1.2.4 Experimental animal studies

More information is needed on the long-term, low-level effects of manganese in order to develop dose-response data. Further studies are also necessary on the neurotoxicity and potential carcinogenicity, teratogenicity, and mutagenicity of manganese and on factors that might affect toxicity such as nutrition, age, disease state, and the presence of other pollutants.

Not enough is known about the essentiality of manganese as a nutrient and more studies are needed on the biochemical role of this metal to obtain a better understanding of toxic mechanisms and to develop a rational basis for the treatment of manganese intoxication.

1.2.5 Epidemiological and clinical studies in man

Studies are required to elucidate the dose-effect and dose-response characteristics of manganese with particular emphasis on the effects of long-term, low-level, inhalation exposure on the respiratory and central nervous systems. Interactions with other pollutants, diet, age, and general health status should be studied in more detail. The effects of manganese on the cardiovascular system, particularly its effects on blood pressure and the myocardium, need to be more fully understood. Reliable diagnostic procedures for manganese intoxication should be established, paying particular attention to the development of methods for its early detection. Additional studies are necessary to assess the embryotoxic potential of manganese and its compounds in communities exposed to elevated levels of manganese in air. Organomanganese compounds may come into widespread use as fuel additives. This would result in increased exposure of the general population to manganese and probably to other combustion products of the additive. Thus, the potential hazards to public health of the use of organomanganese fuel additives should be examined by means of carefully conducted controlled and epidemiological studies.

2. PROPERTIES AND ANALYTICAL METHODS

2.1 Chemical and Physical Properties of Manganese and its Compounds

Manganese, Mn (atomic number $Z = 25$; relative atomic mass $A_r = 54.938$) is an element of the VIIb group of the periodic table of elements, together with technetium and rhenium. It belongs to the first series of d-block transition elements which also contains titanium, vanadium, chromium, nickel, and copper. Because of their electron configuration, transition elements have some characteristic properties: they are all metals; they exist in a variety of oxidation states; and they form many coloured and paramagnetic compounds. Several transition elements have an important role in biological systems.

In the elemental state, manganese is a white-grey, brittle, and reactive metal with a melting point of 1244 °C and a boiling point of 1962 °C. It is the most common transition metal after iron and titanium. It can form compounds in a number of oxidation states, the most important being $+2$, $+3$, and $+7$.

Manganous (Manganese(II), Mn^{2+}) salts are mostly water-soluble, with the exception of the phosphate and carbonate, the solubilities of which are rather low. Dihalides of manganese include MnF_2, $MnCl_2$, $MnBr_2$, and MnI_2. Addition of OH^- ion to the Mn^{2+} solutions gives the gelatinous white hydroxide $Mn(OH)_2$. MnO and MnS are also known. The Mn^{II} complexes are generally weakly coloured (pale pink). Mn^{2+} is in many ways similar to Mg^{2+}, and can replace it in some biological molecules.

Mn_3O_4 (hausmannite) contains both Mn^{II} and Mn^{III}, i.e., $Mn^{II} Mn^{III}_2O_4$. The manganic Mn(III) ion (Mn^{3+}) easily hydrolyses in weak acid solutions into Mn^{2+} and MnO_2. Manganese(III) and manganese(IV) complexes seem to be important in photosynthesis.

Manganese dioxide (MnO_2), found naturally as pyrolusite, is the most important manganese (II) compound. It is insoluble in water and in cold acids. The little-known manganese(IV) ion occurs in blue "hypomangamates".

Manganese(VI) exists in the deep green manganate ion, MnO_4^{2-}, which is stable only in very basic solutions. Otherwise, it breaks down to give the permanganate ion MnO_4^- and MnO_2. The permanganate ion is the best known form of Mn^{VII}. Permanganate, which is a good oxidant in basic solutions, is reduced to Mn^{2+} in acid solutions.

The properties of some inorganic manganese compounds are summarized in Table 1.

Manganese may form a variety of complexes particularly in the $+2$ state. The $+1$ state is present in hexacyano complexes such as $K_5Mn(CN)_6$, which exist also with manganese in the $+3$ state, $K_3Mn(CN)_6$.

Manganese forms various organometallic compounds such as $Mn_2(CO)_{10}$, sodium pentacarbonylmanganate ($NaMn\ (CO)_5$), and manganocene ($(C_5H_5)_2Mn$. However, of major practical interest is methyl-cyclopentadienyl manganese tricarbonyl ($CH_3C_5H_4Mn(CO)_3$), often referred to as MMT, Cl-2 or Ak-33X (antiknock 33X), which has

Table 1. Chemical and physical properties of manganese and some manganese compounds [a]

Compound	Chemical formula	Relative atomic or molecular mass	Melting point (°C)	Boiling point (°C)	Solubility
Manganese	Mn	54.94	1244	1962	Decomposes in cold and hot water; soluble in dilute acid.
(II) acetate	$Mn(C_2H_3O_2)_2$	173.02			Soluble in cold water (decomposes); soluble in alcohol.
(II) carbonate	$MnCO_3$	114.95	decomposes		Soluble in cold water; soluble in dilute acids.
dichloride	$MnCl_2$	125.84	650	1190	Soluble in cold and hot water, and in alcohol.
(II) nitrate	$Mn(NO_3)_2 \cdot 4H_2O$	251.01	25.8	1294	Soluble in cold and hot water, and in alcohol.
(II, III) oxide	Mn_3O_4	228.81	1705		Soluble in hydrochloric acid.
dioxide	MnO_2	86.94	— 0.535		Soluble in hydrochloric acid.
(III) oxide	Mn_2O_3	157.87	— 0.1080		Soluble in acid.
(II) metasilicate	$MnSiO_3$	131.02	1323		Insoluble in water and hydrochloric acid.
(II) sulfate	$MnSO_4$	151.00	700	850 (decomposes)	Soluble in cold and hot water, and in alcohol.
(III) sulfate	$Mn_2(SO_4)_3$	398.06	160		Decomposes in water, soluble in hydrochloric acid, and dilute sulfuric acid.
(II) sulfide	MnS	87.00	decomposes		Soluble in cold water, dilute acid, and alcohol.
(IV) sulfide	MnS_2	119.07	decomposes		Decomposes in hydrochloric acid.
Potassium permanganate [b]	$KMnO_4$	158.00	decomposes	< 240	Soluble in cold and hot water, in sulfuric acid, alcohol, and acetone. Decomposes in alcohol.

[a] From: Weast (1974).
[b] From: Stokinger (1962).

been used as an additive in fuel oil, as a smoke inhibitor, and as an antiknock additive in petrol, usually as a supplement to tetraethyllead.

2.2 Sampling and Analysis

2.2.1 Collection and preparation of samples

Nonmetallic sampling systems should be used for the collection of environmental materials, and suitable precautions should be taken to avoid contamination during the analytical process.

Filters for ambient air particulates must be chosen with care so that trace amounts of manganese in the filter material do not distort the results. Generally the air sampling techniques chosen will depend on the purpose of the investigation. High-volume air samplers and centripeters are expensive, require power points, and are unsuitable for large-scale monitoring at multiple sites. The use of standard deposit gauges is limited to the collection of particles larger than 5 μm; particles with a smaller diameter are deposited only by impaction. In Japanese studies, a high-volume air sampler is used for suspended particulate matter, and a cyclone type low-volume air sampler for suspended particulate matter with a particle size of 10 μm or less (Environment Agency, Japan, 1972).

Sphagnum moss techniques are useful for comparing fallout in different areas or for studying seasonal variations in one area. Continuous sampling drawing measured air volumes through filter paper, or dry deposition on filter papers protected from the rain combined with rain water collecting, may also be used. According to normal practice in emission studies, sampling for manganese at stationary air pollution sources is carried out isokinetically, using a sampling train that will remove manganese efficiently. In the source sampling method used by the US Environmental Protection Agency (1971), it is possible to analyse the particulates collected in the probe, on the filter, and in the water impingers.

Manganese is emitted in automobile exhaust in the form of particulate matter. Concentrations vary according to the natural manganese levels in the fuel and to the concentration of manganese-containing additives, if present. Exhaust particulates may be collected by total or proportional sampling of the hot exhaust or by proportional sampling of the exhaust mixed with air, which allows cooling and condensation of the compounds of greater relative atomic mass associated with short-time ambient exhaust particulates. The second method provides a more realistic assessment of the mass and composition of the primary exhaust particulates. Collection using this technique can be carried out using a single filter, multiple filter, beta gauge (Dresia & Spohr, 1971), or particulate-size-fractionating devices. Gaseous samples may be collected either by the cold-trap technique or on chromatographic columns.

The following considerations are important in the sampling of water for manganese analysis: (*a*) selection of sampling sites; (*b*) frequency of sampling; (*c*) sampling equipment; and (*d*) sample preparation (Brown et al., 1970). Usually, little or no sample preparation is required but freeze-drying operations can be used.

Aqueous samples should be filtered immediately on collection, using a membrane or other suitable filtering material if differentiation between soluble and particulate phases is to be attempted. Once the particulates are collected on a filter, the analytical problems are similar to those of air analysis. Special precautions are required in the handling and storage of solid and aqueous samples with regard to the choice of equipment and containers.

Because of the extremely low concentration of manganese in some biological tissues and body fluids, contamination of the samples constitutes a major difficulty, a fact which is often overlooked or underestimated. It seems likely that the wide variation in manganese concentrations reported, for instance, in serum (section 6.2.1) can partly be explained by contamination.

Steel equipment is considered unsuitable for tissue biopsy, and quartz or glass knives have been suggested as alternatives; the use of a laser beam has also been discussed (Becker & Maienthal, 1975). Versieck et al. (1973a) reported that the radioactivated Menghini needles used in liver biopsy could cause up to 30% manganese contamination. It has also been suggested that skin-pricking is inferior to venepuncture in the drawing of blood samples because of the possible introduction of tissue manganese into the sample (Papavasiliou & Cotzias, 1961). Single transfer of blood through conventional steel needles has caused serious contamination of samples (Cotzias et al., 1966), and the use of platinum-rhodium alloy needles with Kel-F hubs has been proposed to overcome this problem (Becker & Maienthal, 1975).

A considerable contamination problem may arise in the presence of some anticoagulants. Bethard et al. (1964) reported a manganese concentration in heparin of 3.56 μg/ml whereas acid-citrate-dextrose contained only 0.002 μg/ml. Consequently, when heparin was used as an anticoagulant, the manganese concentration was 0.17 ± 0.03 μg/ml compared with 0.00014 μg/ml when acid-citrate-dextrose was used.

Sampling of hair may be complicated by the fact that manganese is associated with melanin-containing structures, black and brown hair containing much higher concentrations of manganese than white hair (Cotzias et al., 1964).

2.2.2 Separation and concentration

Special procedures are not normally necessary for the separation of manganese from other metals prior to the analysis or concentration of samples. Chromatographic methods for the determination of manganese have been reviewed by Fishbein (1973).

2.2.3 Methods for quantitative determination

2.2.3.1 *Optical spectroscopy*

Trace metals, including manganese, have been determined spectroscopically by a number of research workers. With suitable variations in sample preparation, the available standard spectroscopic methods can be used equally well for mineral ores, air particulates, or biological samples (Cholak & Hubbard, 1960; Tipton, 1963; Angelieva, 1969, 1970, 1971; Bugaeva, 1969; Carlberg et al., 1971; El Alfy et al., 1973; Pépin et al., 1973). The advantages of spectroscopy are that it can be applied to most elements with a satisfactory specificity and sensitivity and that it can be used for the simultaneous determination of several elements (US Environmental Protection Agency, 1972, 1973). Drawbacks of the emission spectroscopic assay include the exacting nature of the method, which necessitates the use of highly qualified personnel, the cost of the instrument, the complexity of the method, and the detection limits, which are too high to detect metals occurring in low concentrations (Thompson et al., 1970).

2.2.3.2 *Atomic absorption spectroscopy*

This is the most commonly used method of determining manganese at present, because the procedure is relatively simple and fast and the sensitivity is high. The application to ambient air samples has been described by Thompson et al. (1970), Begak et al. (1972), and Muradov & Muradova (1972). The method is fairly free from interference except for possible matrix effects, which can generally be avoided. Any silica extracted from glass-fibre filters can cause interference unless removed by the addition of calcium to the solution, prior to analysis (Slavin, 1968). Atomic absorption methods have also been used to determine manganese in water and other materials. Little or no preparation of the sample solution is required (Thompson et al., 1970; Tichý et al., 1971; US Environmental Protection Agency, 1974).

The advantage of flameless atomizers is that the determination can be carried out with high sensitivity using only a small sample. The method was initiated by L'vov (1961) to avoid interference caused by reactions in the flame. However, the precision of the results is not necessarily good since atomizing can easily be altered by various conditions such as the type of the sample which, for instance, may stick to the wall of the boat. These difficulties are especially significant when directly atomizing biological samples. Graphite furnace or carbon rod techniques can be used for the direct analysis of water samples, although matrix interference must be checked for and eliminated. Concentration of fresh water can be achieved simply by evaporation. Other variants have been developed for biological substrates, foodstuffs, soils, and plant materials (Ajemian & Whitman, 1960; Suzuki, 1968; Suzuki et al., 1968; Obelanskaja et al., 1971; Bak et al., 1972; Van Ormer & Purdy, 1973).

An atomic absorption assay using direct aspiration of the sample into the burner has been described for the determination of methyl-cyclopentadienyl manganese tricarbonyl (MMT) in gasoline. The drawback of this method is that it does not discriminate between MMT and other manganese compounds (Bartels & Wilson, 1969).

Atomic absorption methods can be classified, according to the type of sample or sample solution to be applied to the atomizer, into: (a) the direct method, in which the sample or sample solution is used directly; and (b) the solvent-extraction method in which a clean-up and concentration process by solvent extraction is carried out before atomizing. The gross matrix effects of saline waters necessitate a preliminary extraction, which usually entails a concentration procedure. Chelating ion-exchange (Riley & Taylor, 1968) and solvent extraction are also often used (Hasegawa & Ijichi, 1973).

2.2.3.3 *Neutron-activation analysis*

This method has a high specificity and sensitivity for very low concentrations of manganese as well as several other elements (Dams et al., 1970). However, the user must be aware that neutron-activation of biological samples may result in the production of isotopes that interfere with the determination of manganese. Irradiated samples are treated by a chemical separation process with a certain amount of manganese carrier and then determined by γ-spectroscopy. The 1810.7 Kev γ line of ^{56}Mn is measured. This method can be used to check the accuracy of results obtained by other analytical methods and for the determination of manganese at very low concentrations in a small number of samples. It is essential to collect particulate matter on filters that have a very low trace element content (ashless filter paper). Variations of this method have been used for determining manganese concentrations in blood and serum (Cotzias et al., 1966) and in plants (Hatamov et al., 1972).

2.2.3.4 *X-ray fluorescence*

The use of X-ray fluorescence spectroscopy provides a means for the non-destructive analysis of elements in sediments and particulates.

X-ray fluorescence can also be used to determine manganese in solutions, if the sample is prepared by freeze-drying. Birks et al. (1972) have made a complete elemental analysis with high sensitivity in 100 seconds using multichannel analysers with 14—24 crystals. The necessity of distinguishing unreactive, structurally incorporated manganese in particulates and sediments from the more reactive absorbed, biogenic, and hydrogenic phases was discussed in a paper by Chester & Hughes (1967), who proposed a selective acid-leaching technique for this purpose. Manganese in water was determined by Watanabe et al. (1972), using a nickel carrier, with a limit of detection of 0.03 μg. Another method that has been developed for the analysis of various elements including manganese, is proton-induced X-ray emission analysis (Johansson et al., 1975). Manganese in dust samples collected by an impactor was detected at nanogram levels using this method.

2.2.3.5 *Other methods*

The periodate method is the classical wet chemical method of analysing air samples for manganese (American Conference of Governmental Industrial Hygienists, 1958). The advantage of this method is that it can be used in almost any chemical laboratory with relatively simple equipment, but the sensitivity (0.1 mg/ m³) is rather poor in comparison with that of other methods (Peregud & Gernet, 1970). This technique has also been widely used for determining total manganese in the soil but it is considered to give a poor estimate of the manganese available to plants.

The permanganate method is the most commonly used method for the analysis of manganese in water samples. Interference caused by manganese in the glassware has to be eliminated when the manganese level in the sample is low, and prior removal of organic material may be necessary. However, not all forms of manganese likely to occur in water can be measured by the permanganate method (e.g., the complexes of trivalent manganese and manganese dioxide), and an improved, simply performed formaldoxime method has been developed for the analysis of both water and soils (Samohvalov et al., 1971; Cheeseman & Wilson, 1972).

A rapid drop quantitative method, developed for determining manganese in the air of the working environment, is based on the colour reaction of manganese ions with potassium ferricyanide. The method is specific and results compare well with those obtained by emission spectroscopy (Muhtarova et al., 1969).

A kinetic method based on the ability of manganese to catalyse the atmospheric oxidation of the morin-beryllium complex has been developed for the determination of manganese in atmospheric precipitates (Morgen et al., 1972). Polarography can be used for determining manganese in industrial waste waters with a sensitivity of 0.05 mg/litre. Chromium interference can be removed by phosphate precipitation (Bertoglio-Riolo et al., 1972). A similar method for analysing animal foodstuffs, organs, and tissues has been developed for samples weighing only 2 g. Interference caused by iron can be avoided by precipitating it with a mixture of ammonium chloride and ammonium hydroxide (Usovic, 1967). An alternative method for the determination of manganese in biological material is electron paramagnetic resonance spectroscopy (Cohn & Townsend, 1954; Miller et al., 1968).

Spark source mass spectroscopy is probably a suitable method for the determination of manganese in petrol (US Environmental Protection Agency, 1975). Cyclopentadienyl manganese tricarbonyl can be determined by treating the sample with nitric and sulfuric acid and subsequently converting the manganese to permanganate (Byhovskaja et al., 1966).

2.2.3.6 *Comparability of methods*

As already stated, atomic absorption spectroscopy combined, when necessary, with a separation solvent procedure, can be applied to most environmental samples. Each of the other methods described has its particular advantages and characteristics and can be used according to the need for sensitivity and to the type of sample. Studies such as that of Harms (1974) on the comparison of data from several different analytical methods are useful. The intercomparison of analytical techniques carried out under the responsibility of EURATOM [a] also provides interesting information. In this study, good agreement was obtained when neutron-activation analysis, X-ray fluorescence, emission spectroscopy, and atomic absorption spectroscopy were compared for the determination of manganese.

---

[a] EURATOM (unpublished data, 1974) Chemical analysis of airborne particulates: intercomparison and evaluation of analytical techniques. In: Guzzi, G., ed. *Minutes of the Meeting held at Ispra, Italy, 8—9 July 1974*, Ispra Establishment, Chemistry Division, Joint Research Centre of the European Communities, 33 pp.

3. SOURCES OF MANGANESE IN THE ENVIRONMENT

3.1 Natural Occurrence

Manganase is widely distributed in nature but does not occur as the free metal. The most abundant compounds are the oxide (in pyrolusite, brannite, manganite, and hausmannite), sulfide (in manganese blende and hauserite), carbonate (in manganesespar), and the silicate (in tephroite, knebelite, and rhodamite). It also occurs in most iron ores in concentrations ranging from 50—350 g/kg, and in many other minerals throughout the world.

A rough estimate of the average concentration of manganese in the earth's crust is about 1000 mg/kg (NAS-NRC, 1973). Manganese concentrations in igneous rock may range from about 400 mg/kg in low-calcium granitic rock to 1600 mg/kg in ultrabasic rock and sedimentary rocks. Deep sea sediments contain concentrations of about 1000 mg/kg (Turekian & Wedepohl, 1961). It has been reported that the manganese content of coal ranges from 6 to 100 mg/kg (Ruch et al., 1973) and that of crude oil from 0.001 to 0.15 mg/kg (Bryan, 1970).

In soil, manganese concentrations depend primarily on the geothermal characteristics of the soil, but also on the environmental transformation of natural manganese compounds, the activity of soil microorganisms, and the uptake by plants.

Although the principal ores are only slightly soluble in water, gradual weathering and conversion to soluble salts contribute to the manganese contents of river and sea water. Considerable amounts of manganese are present in deposits in large areas of the oceans in the form of nodules. These are formed continuously at a rate of several million tonnes per year (Schroeder et al., 1966). The average concentration of manganese in these nodules is about 200 mg/kg (Zajic, 1969) with a range of about 150—500 mg/kg (Schroeder et al., 1966).

3.2 Industrial Production and Consumption

Elemental manganese was isolated in 1774, though the oxide has been used in the manufacture of glass since antiquity. The total world production of manganese, which was 18 million tonnes in 1969, rose to about 27 million tonnes in 1975. However, consumption, which had risen by 20% between 1970 and 1975, dropped by 3% in 1975 (Mineral Yearbook, 1975, 1977).

Fumes, dust, and aerosols from metallurgical processing, mining operations, steel casting (Mihajlov, 1969) and metal welding and

Table 2. Emission factors for manganese

Mining	0.09 kg/tonne of manganese mined
Processing	
manganese metal	11.36 kg/tonne of manganese processed
ferromanganese	
blast furnace	1.86 kg/tonne of ferromanganese produced
electric furnace	10.86 kg/tonne of ferromanganese produced
silicomanganese	
electric furnace	31.55 kg/tonne of silicomanganese produced
Reprocessing	
carbon steel	
blast furnace	10.22 kg/1000 tonnes of pig iron produced
open-hearthfurnace	23.18 kg/1000 tonnes of steel produced
basic oxygen furnace	20.00 kg/1000 tonnes of steel produced
electric furnace	35.45 kg/1000 tonnes of steel produced
cast iron	150.00 kg/1000 tonnes of cast iron
welding rods	7.27 kg/tonne of manganese processed
nonferrous alloys	5.45 kg/tonne of manganese processed
batteries	4.54 kg/tonne of manganese processed
chemicals	4.54 kg/tonne of manganese processed
Consumer uses	
coal	3.50 kg/tonne of coal burned

From: Davis & Associates (1971).

cutting, (Erman, 1972), mainly in the form of manganese oxide are the principal sources of environmental pollution. Emissions into the atmosphere from blast and electric furnaces vary considerably depending on the process involved and the degree of control exercised. Dust from the handling of raw materials in metallurgical processing and other manufacturing activities probably makes only a small contribution to the atmospheric concentration of manganese. Calculated emission factors for manganese are given in Table 2.

3.2.1 Uses

Over 90% of the manganese produced in the world is used in in the making of steel, either as ferromanganese, silicomanganese, or spiegeleisen. Manganese is also used in the production of nonferrous alloys, such as manganese bronze, for machinery requiring high strength and resistance to sea water, and in alloys with copper, nickel, or both in the electrical industry. In dry-cell batteries, manganese is used in the form of manganese dioxide, which is also used as an oxidizing agent in the chemical industry. Many manganese chemicals, eg., potassium permanganate, manganese(II) sulfate, manganese dichloride, and manganese dioxide are used in fertilizers, animal feeds, pharmaceutical products, dyes, paint dryers, catalysts, wood preservatives and, in small quantities, in glass and ceramics. Some of these uses contribute to environmental pollution.

3.2.2 Contamination by waste disposal

The disposal of liquid and solid waste products containing manganese may contribute to the contamination of land, water courses, and soil. For example, sludges and various waste waters containing manganese are used in the production of micronutrient fertilizers (Eliseeva, 1973) and manganese slurries have been used in the production of clay blocks for road construction. Information concerning the degree of pollution arising from the incineration of refuse containing manganese is not available.

3.2.3 Other sources of pollution

The emission of manganese from motor vehicles powered by petrol that does not contain manganese additives has been estimated to average 0.03—0.1 mg/km (Moran et al., 1972; Gentel et al., 1974a; Gentel et al., 1974b).

Methylcyclopentadienyl manganese tricarbonyl (MMT) was initially marketed in the USA as a supplement to tetraethyl lead in an antiknock preparation. During the 1960s, it was introduced as a fuel-oil combustion improver and as a smoke suppressant for gas turbines using liquid fuels. In 1974, it came into commercial use as a fuel additive in unleaded petrol in the USA; in 1976 about 20% of the fuel was unleaded, and 40% of this amount contained MMT at an average concentration of 10.56 mg/litre (0.04 g/US gallon) (Ethyl Corporation, private communication). The use of MMT is likely to increase during the coming years. At the manufacturer's recommended maximum level of MMT (a manganese concentration of 33 mg/litre),[a] the emission of MMT is approximately 0.62—3.1 μg/km (1—5 μg/mile); levels of about 0.62—1.55 μg/km (1—2.5 μg/mile) have been reported in lubrication oil (Hurn et al., 1974). This low emission rate together with the fact that MMT rapidly undergoes photochemical decomposition (section 5.8) suggests that exposure to the parent compound through the exhaust gas would be low.

Taking data on lead emissions in exhaust gas as a model, it has been calculated that the use of MMT in petrol might result in the emission of 0—0.25 μg of manganese per m³ of air, with a median of 0.05 μg/m³, and that the organic component of this would be about 1.2×10^{-5} μg/m³ (Ter Haar et al., 1975). This is not far from the estimate of 0.05—0.2 μg/m³ made by Keane & Fisher, (1968). It has been reported that 50% of emitted manganese particles have a mass median diameter (MMD) of 0.5 μm or less (Moran, 1975).

---

[a] In June 1977, the manufacturer reduced the recommended maximum level of manganese in petrol to 16 mg/litre, bringing about a corresponding cut in the estimated emission levels (Ethyl Corporation, private communition).

At the 1975 SAE Automobile Engineering Meeting, it was claimed that the use of manganese in petrol resulted in increased total particulate emissions that could not be totally accounted for on the basis of increased manganese content (Moran, 1975). This was disputed at the same meeting by Desmond (1975), who argued that the figures presented by Moran (1975) for increased total particulate emissions were compatible with the theoretical maximum emissions of Mn_3O_4 resulting from combustion of the manganese in the fuel.

It appears that the use of MMT in petrol causes increased emission of hydrocarbons (Gentel et al., 1974b; Hurn et al., 1974; Kocmond et al., 1975). However, there is no conclusive evidence to indicate that MMT decreases the efficiency of catalysts (Faggan et al., 1975; Moran, 1975).

It is possible that MMT in petrol increases aldehyde emissions, though the data so far available are conflicting (Ethyl Corporation, 1974; Gentel et al., 1974b; Hurn et al., 1974). Too little information is available to draw any conclusions with regard to the effects of MMT in petrol on the emission of polynuclear aromatic hydrocarbons. Tests performed by the Ethyl Corporation (1974) showed a decrease in benzo(a)pyrene concentrations in exhaust gas. A similar decrease in benzo(a)pyrene concentrations was reported by Lerner (1974) using an analogous compound, cyclopentadienyl manganese tricarbonyl. In one study, it was shown that MMT in petrol could decrease atmospheric visibility (Kocmond et al., 1975). Results of other studies conducted by the Ethyl Corporation (1971) indicated that comparatively high concentrations of manganese in air were needed to influence the reaction converting sulfur dioxide to sulfuric acid and sulfates. Thus, the reaction rate was unchanged at a manganese concentration of 4 $\mu g/m^3$ and no effect was detectable at a concentration of 36 $\mu g/m^3$, when the humidity was below 70%.

The effects of MMT in petrol on the emission of carbon monoxide and oxides of nitrogen are not clear (Moran, 1975).

Another organic manganese compound, manganese ethylene-bis-dithiocarbamate (Maneb), is used as a fungicide.

A large-scale investigation was made in Japan using a pilot plant, equipped with a desulfurization device containing activated manganese dioxide, to explore its influence on manganese levels in the surrounding environment. The operation of the device increased the manganese level in air by an average value of 0.002 $\mu g/m^3$ (Ministry of International Trade and Industry & Ministry of Health and Welfare, 1969).

Minor uses of manganese compounds in the manufacture of linoleum and calico printing and in the manufacture of matches and fireworks may be an additional source of environmental contamination.

4. ENVIRONMENTAL LEVELS AND EXPOSURE

4.1 Air

4.1.1 Ambient air

The natural level of manganese in air is low. A concentration in air of 0.006 μg/m³ at a height of 2500 m and an annual average concentration of 0.027 μg/m³ at 823 m were reported by Georgii et al. (1974). In rural areas, manganese levels in air may range from 0.01 to 0.03 μg/m³ (US Environmental Protection Agency, 1973).

Because nearly all the manganese emitted into the atmosphere is in association with small particles, it may be distributed over considerable distances. According to Lee et al. (1972), about 80% of manganese emitted into the atmosphere is associated with particles with a mass median equivalent diameter of less than 5 μm and about 50% with particles of less than 2 μm. Thus, most of the particles are within the respirable range.

A survey of manganese concentrations in suspended particulate matter, conducted during the period 1957—1969 at some 300 urban and 300 nonurban sites in the USA, has been summarized by the US Environmental Protection Agency (1975). Annual average manganese concentrations ranged from less than 0.099 μg/m³ for about 80% of the sites to more than 0.3 μg/m³ for about 5% of the sites (Table 3). In areas associated with local ferromanganese or silico-

Table 3. Number of National Air Surveillance Network (NASN) stations within selected annual average manganese concentration intervals, 1957—1969 [a]

Year		Concentration interval (μg/m³)				Total
		< 0.099	0.100—0.199	0.200—0.299	> 0.300	
1957—1963	No. stations	76	29	10	13	128
	%	59.4	22.7	7.8	10.2	100
1964	No. stations	68	12	6	7	93
	%	73.1	12.9	6.5	7.5	100
1965	No. stations	132	14	5	6	157
	%	84.1	8.9	3.2	3.8	100
1966	No. stations	113	8	4	3	128
	%	88.3	6.3	3.1	2.3	100
1967	No. stations	121	13	4	4	142
	%	85.2	9.2	2.8	2.8	100
1968	No. stations	126	11	2	6	145
	%	86.9	7.6	1.4	4.1	100
1969	No. stations	169	23	9	8	209
	%	80.9	11.0	4.3	3.8	100
1957—1969	No. stations	805	110	40	47	1002
	%	80.4	11.0	4.0	4.7	100

[a] From: US Environmental Protection Agency (1975).

Table 4. National Air Surveillance Network (NASN) stations with annual average manganese concentrations greater than 0.5 μg/m³ [a]

Year	Station	Manganese concentration (μg/m³)		
		Average	Max. quarterly	Max. 24-h
1958	Charleston, W.VA	0.61	1.10	7.10
1959	Johnstown, PA	2.50	5.40	7.80
	Canton, OH	0.72	1.10	2.20
1960	Gary, Ind.	0.97		3.10
1961	Canton, OH	0.57		2.90
	Philadelphia, PA	0.70		> 10.00
1963	Johnstown, PA	1.44		6.90
	Philadelphia, PA	0.62		3.70
1964	Charleston, W.VA	1.33		> 10.00
1965	Johnstown, PA	2.45	3.90	
	Philadelphia, PA	0.72	1.70	
	Lynchburg, VA	1.71	2.50	
	Charleston, W.VA	0.60	1.70	
1966	Niagara Falls, NY	0.66	1.30	
1967	Knoxville, TN	0.81	1.50	
1968	Johnstown, PA	3.27		14.00
1969	Niagara Falls, NY	0.66	1.30	
	Johnstown, PA	1.77	2.10	
	Philadelphia, PA	0.50	1.30	

[a] From: US Environmental Protection Agency (1975).

manganese industries such as Johnstown, Charleston, and Niagara Falls, the annual average concentrations ranged upwards from 0.50 μg/m³ (Table 4). The average 24-h concentrations in such places can exceed 10 μg/m³ and may present an important health risk. Urban centres without major foundry facilities, such as New York, Los Angeles, and Chicago, exhibited annual average manganese concentrations in air ranging from 0.03 to 0.07 μg/m³, whereas in cities with these facilities, such as Pittsburg, Birmingham, and East Chicago, values ranged from 0.22 to 0.30 μg/m³ (US Environmental Protection Agency, 1973). These concentrations are in agreement with those found in other studies from the USA (Brar et al., 1970; Lee et al., 1972). The highest reported annual average concentration of 8.3 μg/m³, was measured in Kanawha Valley, West Virginia, during 1964—65. The major source of pollution was a ferromanganese plant situated in a nearby area (US Environmental Protection Agency, 1975).

Manganese values from air sampling sites in the United Kingdom during 1971—1972 ranged from 0.004 to 0.049 μg/m³; Keane & Fisher (1968) reported mean manganese concentrations of 0.013—0.033 μg/m³ in relatively unpolluted areas of the United Kingdom.

In the Federal Republic of Germany, manganese concentrations were found to range from 0.08 to 0.16 μg/m³ in different areas of Frankfurt, with a maximum 24-h concentration of 0.49 μg/m³ (Georgii & Müller, 1974), whereas in a residential area of Munich levels of 0.030—0.034 μg/m³ were reported, with 0.06—0.27 μg/m³ in a street with heavy traffic (Bouquiaux, 1974).

The Environment Agency, Japan (1975) reported an annual mean manganese concentration in the air of Japanese cities of about 0.02—0.80 μg/m³ with maximum 24-h concentrations of 2—3 μg/m³ (Environment Agency, Japan, 1975). Studies are also available from a district in Kanazawa, Japan, close to a plant using electric furnaces for the production of manganese alloys. Average levels during 1970 varied from 1.1 to 9.8 μg/m³, when measured over 2-day periods at a point 300 m from the emitting source. Unpolluted areas of the same city showed average levels of 0.035 μg/m³ during the period 1968—1970 (Itakura & Tajima, 1972). When manganese concentrations were measured at underground shopping districts adjoining subway stations in Tokyo, Osaka, and Nagoya, open-air concentrations of 0.042—0.074 μg/m³ and subway concentrations of 0.040—0.353 μg/m³ were found, indicating that heavy subway traffic on railway lines containing manganese as a ferroalloy may increase manganese exposure (Japan Environmental Sanitation Centre, 1974).

Thus, it can be concluded that annual average levels for manganese in ambient air in nonpolluted areas range from approximately 0.01 to 0.03 μg/m³, while in urban and rural areas without significant manganese pollution, annual averages are mainly in the range of 0.01—0.07 μg/m³. With local pollution near foundries, this level can rise to an annual average of 0.2—0.3 μg/m³ and in the presence of ferro- and silicomanganese industries, to over 0.5 μg/m³. The data available are not adequate for drawing valid conclusions with respect to trends in ambient manganese concentrations.

4.1.2 Air in workplaces

In recent years, most of the industrialized countries have established occupational exposure limits for manganese. Thus, working conditions have improved and earlier reports of excessive exposure to manganese do not always represent more recent conditions. This should be borne in mind when considering the information presented in this section.

According to one report (Ansola et al., 1944a), Chilean manganese miners were exposed to manganese concentrations in air of 62.5—250 mg/m³. However in a later study in a Chilean mine, Schuler et al. (1957) reported a concentration range of 0.5—46 mg/m³, the highest levels being found in connection with the drilling of pure, dry ore and the drilling of manganese-bearing rock. Manganese concentrations of up to 926 mg/m³ of air were found in Moroccan mines (Rodier, 1955). Flinn et al. (1940) recorded a manganese concentration of 173 mg/m³ in an ore-crushing mill in the USA but a much later survey of dust levels in the air of a ferromanganese crushing plant in the United Kingdom (as measured by personal sampling devices) showed manganese concentrations of 0.8—8.6 mg/m³. The device of one man cleaning down the crusher showed an exceptionally high concentration of 44.1 mg/m³. When levels in air were measured at fixed sampling points, they ranged from 8.6 to 83.4 mg/m³ (Department of Health & Social Security, unpublished data).[a]

In an electric steel foundry in Japan, manganese concentrations ranged from 4.0 to 38.2 mg/m³ around an electric furnace and from 4.9 to 10.6 mg/m³ around the mouth of the kiln (Ueno & Ohara, 1958).

In studies in the USSR reported by Mihajlov (1969), manganese concentrations in air of 0.3 mg/m³ or more were found in 98% of 1905 samples collected in the furnace area of a steel shop, during the period 1948—1963. The levels reached 1.8—2.4 mg/m³ during melting operations and increased to as much as 10 mg/m³, when the molten steel was being poured. Additional data on manganese concentrations in air can be found in section 9.1.

Few studies have included details of the size distribution of manganese dust, which is of importance in the evaluation of dust absorption following inhalation. Akselsson et al. (1975) reported manganese concentrations of up to 3 mg/m³ in the breathing zone of welders. The highest concentrations were associated with particles ranging in size from 0.1 to 1.0 μm. This is in agreement with the finding that 80% of particles from a ferromanganese furnace ranged in size from 0.1 to 1.0 μm (Sullivan, 1969). In studies by Smyth et al. (1973), more than 99% of the particles in airborne fume around a blast furnace were smaller than 2 μm and 95% of airborne dust particles at a crushing and screening plant were smaller than 5 μm.

4.2 Water

Manganese may be present in fresh water in both soluble and suspended forms. However, in most reported studies, only total manganese has been determined.

Surface waters of various American lakes were found to contain from 0.02 to 87.5 μg of manganese per litre with a mean of 3.8 μg/litre (Kleinkopf, 1960). In two other studies the contents of large rivers in the USA ranged from below the detection limit to 185 μg/litre (Durum & Haffty, 1961; Kroner & Kopp, 1965). A range of 0.8—28.0 μg/litre was found in Welsh rivers (Abdullah & Royle, 1972). Manganese concentrations at 37 river sampling sites in the

---

[a] Department of Health and Social Security (1975) *Environmental health criteria for manganese and its compounds: Review of work in the United Kingdom, 1967—1973.*

United Kingdom (Department of Health and Social Security, 1975
— unpublished) and in the Rhine and the Maas and their tributaries
(Bouquiaux, 1974) ranged from 1 to 530 μg/litre. There are some
reports indicating a seasonal variation in the manganese contents of
rivers (Beščetnova et al., 1968; Kolesnikova et al., 1973) and inshore
waters, manganese levels being lowest during the winter months
(Morris, 1974). High manganese concentrations reaching several
mg/litre have been found in waters draining mineralized areas
(Kolomijeeva, 1970; Department of Health and Social Security, 1975
— unpublished) and in water contaminated by industrial discharges
(Kozuka et al., 1971).

In the USSR, groundwater not associated with manganese-bearing
rock, contained manganese concentrations ranging from 1 to 250
μg/litre (Kolomijeeva, 1970). A comparatively high average concentration of 0.55 mg/litre was reported in a study of 6329 untreated
samples of groundwater in Japan (Kimura et al., 1969) and concentrations ranging from 0.22 to 2.76 mg/litre were found in deep well
water in the Takamatsu City area (Itoyama, 1971).

An average concentration of manganese in seawater of 0.4
μg/litre was reported by Turekian (1969). In other studies on the
manganese contents of sea water in the North Sea, the Northeast
Atlantic, the English Channel, and the Indian Ocean, concentrations
ranged from 0.03 to 4.0 μg/litre with mean values of 0.06—1.2
μg/litre. In estuarine and coastal waters of the Irish Sea and in
waters along the North Sea shores of the United Kingdom, values
ranging from 0.2 to 25.5 μg/litre have been reported with mean
values of 1.5—6.1 μg/litre (Topping, 1969; Preston et al., 1972; Jones
et al., 1973; Bouquiaux, 1974).

Manganese concentrations in treated drinking-water supplies in
100 large cities in the USA ranged from undetectable to 1.1 mg/litre,
with a median level of 5 μg/litre; 97% of the supplies contained concentrations below 100 μg/litre (Durfor & Becker, 1964). According
to a US Public Health Service survey quoted by Schroeder (1966),
manganese levels in tap water from 148 municipal supplies ranged
from 0.002 to 1.0 mg/litre, with a median level of 10 μg/litre. Mean
concentrations of manganese in drinking-water in the Federal
Republic of Germany were reported to range from 1 to 63 μg/litre
(Bouquiaux, 1974).

4.3 Soil

The average concentration of manganese in soils is probably
about 500—900 mg/kg (NAS/NRC, 1973). Earlier analyses are of
doubtful value, as errors arising from contamination and interference
with other substances were not fully appreciated (Mitchell, 1964).
The significance of manganese levels in soils depends largely on the

type of compounds present and on the characteristics of the soil such as the pH and the redox potential. Accumulation usually occurs in the subsoil and not in the surface, 60—90% of manganese being found in the sand fraction of the soil. In well-drained areas, the manganese contents of stream sediments and of parent rocks and soils have been found to be comparable. In areas of poorly-drained, peaty gleys and podzols, stream sediments may be greatly enriched. For example, stream sediments from poorly drained Welsh moorlands with rock and soil concentrations of 540 mg/kg and 300 mg/kg, respectively, contained an excess of 1% manganese (Nichol et al., 1967).

Soddy-podzolic soils in the USSR contained manganese concentrations of 21—200 mg/kg, chernozem soils, up to 6400 mg/kg, and boggy soils, 10—500 mg/kg. Mobile manganese in the USSR soils varied from 23 to 149 mg/kg (Vasilevskaja & Bogatyrev, 1970). In Belgium, loess formation in a forest region contained manganese concentrations of 113—450 mg/kg. In a semi-industrialized region, concentrations ranging from 135 to 320 mg/kg were found, while in sandy uncultivated soil, concentrations ranged from 30 to 43 mg/kg (Bouquiaux, 1974).

4.4 Food

The manganese contents of various foodstuffs vary markedly (Table 5).

Table 5. Manganese levels in some foodstuffs

Category	Manganese (mg/kg wet weight)	
	Shroeder et al. (1966)	Guthrie (1975)
Cereals		
barley, meal	17.8	9.9
corn	2.1	3.8
rice, polished	1.5	9.6
unpolished	2.1	32.5
rye	13.3	34.6
wheat	5.2—11.3	13.7—40.3
Meat and poultry	< 0.1—0.8	< 0.1—2.7
Fish	≤ 0.1	0.1—0.5
Dairy products		
milk	0.2	0.5
butter	1.0	0.1
Eggs	0.5	0.3
Vegetables		
beans	0.2	1.8
peas	0.6	2.6
cabbage	1.1	0.8
spinach	7.8	1.8
tomatoes	0.3	0.2—0.6
Fruit		
apples	0.3	0.2—0.3
oranges	0.4	0.3
pears	0.3	0.1—0.4
Nuts		
walnuts	7.5	19.7

In cereal crops from the USSR, manganese concentrations varied from 2 to 100 mg/kg wet weight, concentrations in pulse crops ranged from 0.36 to 32 mg/kg, and those in root crops from 0.2 to 15 mg/kg; beet crops contained up to 37 mg/kg (Aljab'ev & Dmitrienko, 1971; Musaeva & Kozlova, 1973).

The edible muscle tissue of 8 common commercial species of fish in New Zealand was reported by Brooks & Rumsey (1974) to have mean concentrations of manganese ranging from 0.08 to 1.15 mg/kg wet weight. Similar values (0.03—0.2 mg/kg wet weight) were found in North Sea fish. In cod and plaice, most values were lower than 0.1 mg/kg. Shellfish may concentrate manganese. Scallops, oysters, and mussels dredged from Tasman Bay contained average manganese levels of 111 mg, 8 mg, and 27 mg/kg dry weight, respectively (Brooks & Rumsey, 1965).

High concentrations of manganese have been found in tea including levels of 780—930 mg/kg in the finished leaves (Nakamura & Osada, 1957) and 1.4—3.6 mg/litre in liquid tea (Nakagawa, 1968).

In most human studies, the average daily intake of manganese, via food, by an adult has been reported to be between 2 and 9 mg/day. Values of about 2.3—2.4 mg/day have been reported from the Netherlands (Belz, 1960) and the USA (Schroeder et al., 1966). North et al. (1960) obtained an average daily intake of 3.7 mg for 9 American college women, and Tipton et al. (1969), using the duplicate portion method, reported 50-week, mean daily intakes of 3.3 and 5.5 mg, respectively, for two American adult males. Similarly, an average intake of 4.1 mg/day was reported from a Canadian composite diet (Méranger & Smith, 1972). In a study by Soman et al. (1969), also using the duplicate portion method, the average manganese intake for Indian adults was 8.3 mg/day, while the intake from drinking-water ranged from 0.004 to 0.24 mg/day. These results agree well with previously reported values for Indian adults on a rice diet (9.81 mg of manganese/day) and on a wheat diet (9.61 mg of manganese/day) (De, 1949).

The daily intake of manganese by bottlefed and breastfed infants is very low because of the low concentrations of manganese in cow's milk and, especially, in breast milk (McLeod & Robinson, 1972a). Widdowson (1969) reported a daily intake of 0.002 mg/kg body weight for 1-week-old babies. Values of a similar order of magnitude (0.002—0.004 mg/kg) have been reported for the first 3 months of life by Belz (1960) and McLeod & Robinson (1972a). When a child is established on a mixed food regimen after 3—4 months of age, the intake increases considerably (McLeod & Robinson, 1972a).

Belz (1960) reported a daily intake of 1.7 mg for children aged 7—9 years, and Schlage & Wortberg (1972) reported intakes of 1.4 mg/day for 6 children aged 3—5 years, and 2.2 mg/day for 5 children aged 9—13 years, corresponding to 0.08 mg and 0.06 mg/kg body weight, respectively. Day-to-day intake varied considerably, the maximum intake being 10 times the minimum. Similar values for daily intake were obtained by Alexander et al. (1974) for 8 children aged between 3 months and 8 years; the mean intake was 0.06 mg/kg body weight.

4.5 Total Exposure from Environmental Media

Based on annual average air concentrations and a respiratory rate of 20 m³/day, an estimate of the daily exposure to manganese of populations living in areas without manganese-emitting industries would be less than 2 μg/day. For populations living in areas with major foundry facilities, the value is likely to be about 4—6 μg, while in areas associated with ferromanganese or silicomanganese industries, the exposure may rise to 10 μg, and 24-peak values may exceed 200 μg.

Considering the manganese concentrations in the vast majority of drinking-water supplies, and assuming a water intake of 2 litres per day, the average daily intake of manganese with drinking-water would be about 10—50 μg with a range of about 2—200 μg. Although the variation is considerable, an intake exceeding 1.0 mg/day would be exceptional.

The daily intake of manganese from food appears to be 2—9 mg. Some European and American studies suggest a likely range of 2—5 mg, while in countries where grain and rice make up a major portion of the diet, the intake is more likely to be in the range of 5—9 mg. The consumption of tea may substantially add to the daily intake.

The average intake for children from a very early age up to adolescence is about 0.06—0.08 mg/kg body weight whereas for breastfed or bottlefed infants intake is only about 0.002—0.004 mg/kg body weight.

5. TRANSPORT AND DISTRIBUTION IN ENVIRONMENTAL MEDIA

5.1 Photochemical and Thermal Reactions in the Lower Atmosphere

Atmospheric manganese compounds seem to promote the conversion of sulfur dioxide to sulfuric acid (Coughanowr & Krause, 1965; Matteson et al., 1969; Ethyl Corporation, 1971; McKay, 1971). However, the concentration of manganese required to achieve this conversion and the significance of its effect remain unknown. The available evidence seems to indicate that a higher concentration of atmospheric manganese than is normally observed would be necessary.

Manganese dioxide reacts with nitrogen dioxide, in the laboratory, to form manganous nitrate (Schroeder, 1970). There is the possibility that such a reaction might occur in the atmosphere but further studies are needed before any conclusion can be reached.

5.2 Decomposition in Fresh Water and Seawater

All water contains manganese derived from soil and rocks. Manganese in seawater is found mostly as manganese dioxide (MnO_2), some of which is produced from manganese salts by several species of bacteria common to soils and ocean muds. The aqueous chemistry of manganese is complex. Mobilization of manganese is favoured by low Eh and/or pH conditions. Thus acid mine-drainage waters can give rise to high environmental concentrations of dissolved manganese. Mitchell (1971) showed that mobilization was greatly enhanced in acid, poorly drained podzolic soils and groundwaters. It was suggested by Nichol et al., (1967) that, in acid waterlogged soils, manganese passes freely into solution and circulates in the groundwaters but that it is precipitated on entering stream waters with average pH and Eh, thus giving rise to stream sediments enriched with manganese.

Particulate material suspended in natural waters may contain an appreciable proportion of manganese. Preston et al., (1972) found that 67—84% of the total manganese in shoreline and offshore areas of the British Isles was associated with particulate matter that contained manganese levels of several hundred mg/kg. Levels of particulate manganese present in ocean waters are low in comparison with levels of dissolved manganese. However, much larger amounts of particulate manganese occur in estuarine and river waters, where resuspension of bottom material may occur. Spencer & Sachs (1970) found that organic particulate matter in the Gulf of Maine was predominantly regenerated in the water column and that the amount of manganese transported to the sediments in this way was negligible.

In deep-sea sediments, manganese is concentrated in the form of both crustal material and coastal and shelf sediments. The composition of manganese nodules on the ocean floors is related to factors such as water composition, sedimentation rates, volcanic influences, and organic productivity. Regional variations have also been observed, especially in the Atlantic Ocean (Elderfield, 1972).

5.3 Atmospheric Washout and Rainfall

On the basis of samples taken at 32 stations in the USA, Lazrus et al. (1970) concluded that the manganese in atmospheric precipitation was derived mainly from human activity. The average manganese concentration in the samples was 0.012 mg/kg. These data do not show the immediate influence of major sources of industrial emissions.

5.4 Run-off into Fresh Water and Seawater

Aerosols, pesticides, limestone and phosphate fertilizers, manures, sewage sludge, and mine wastes have all been identified as possible sources of soil contamination that can add to the manganese burden of fresh water and seawater (Lagerwerff, 1967). The concentrations of trace elements in soil additives are generally low and do not significantly affect the total manganese content of soil (Swaine, 1962; Mitchell, 1971).

5.5 Microbiological Utilization in Soils

Manganese cycles in the soil have been proposed involving di-, tri-, and tetravalent manganese. Divalent manganese is transformed through biological oxidation to the less available trivalent form and later, through dismutation, the Mn^{+++} form is biologically reduced to Mn^{++}. A dynamic equilibrium may exist between all forms. The oxidizing power of higher oxides increases with acidity and thus reduction by organic matter is more likely at low pH values. If the oxygen tension is low, biological reduction can take place at any pH value. Bacterial oxidation is very slow or absent in very acid soils and Mn^{++} predominates; organic matter can reduce the higher oxides. In alkaline soils, the divalent form nearly disappears: bacterial oxidation is rapid and reduction by organic matter is slow. In well-aerated soils with a pH of more than 5.5, soil microorganisms can oxidize the divalent form rapidly. The rates of exchange between the various forms are not known at the present time but there is a very pronounced seasonal variation. This is probably due to oxidation and reduction induced by microbial action. The manganous form predominates in summer and the manganic form in winter, though the opposite is said to be true for alkaline soils (Zajic, 1969).

5.6 Uptake by Soil and Plants

It appears that plants mainly absorb manganese in the divalent state and that the availability of soil manganese is closely influenced by the activity of microorganisms that can alter pH and oxidation reduction potentials. Reducing the soil pH or the soil aeration by flooding or compaction favours the reduction of manganese to the Mn^{++} form and thereby increases its solubility and availability to plants. Heavy fertilization of acid soils without liming (particularly with materials containing chlorides, nitrates, or sulfates) may also increase manganese solubility and availability. Under some conditions of pH and aeration, the addition of organic compounds to soil can increase the chemical reduction of manganese and its uptake by plants. In a study by NAS/NRC (1973), it was shown that the capacity of plants to absorb manganese varied according to species. For example, in 20 different species of flowering plants, the absorption capacity of some species was 20—60 times greater than that of the species with the lowest capacity for absorbing the element (NAS/NRC, 1973).

Areas with low manganese concentrations in the soil (below 500 mg/kg) are associated with low manganese levels in the herbage (30—70 mg/kg dry weight) (Department of Health & Social Security, 1975 — unpublished). Liming has been shown to reduce the availability of manganese in soils; on plots with pH values ranging from 5.0 to 7.0, the average manganese content of clover fell from 55 to 12 mg/kg and that of rye grass from 104 to 13 mg/kg, after liming (Reith, 1970). Nitrogen applications consistently reduce the availability of manganese. Organic material associated with a high pH can produce organic complexes of divalent manganese leading to insufficient available manganese for susceptible plants such as peas or cereals. Aging of manganese oxides reduces their availability. Manganese toxicity in plants may occur in soils containing manganese levels exceeding 1000 mg/kg dry weight; this generally occurs in very acid soils and can usually be remedied by liming (Mitchell, 1971). It should be noted that the total manganese content of soil is of little biological significance, since only a small amount is present in an available form.

The uptake of manganese by barley plants is stimulated by the presence of microorganisms, which also appear to break down EDTA-manganese chelates (Barber & Lee, 1974). On a dry-weight basis, perennial rye and timothy grass have been shown to have about three times the manganese content of lucerne, and rather more than tetraploid red clover. Under deficiency conditions, plants destined for herbage contained manganese concentrations of less than 10 mg/kg dry weight (Fleming, 1974).

5.7 Bioconcentration

Terrestrial mammals may concentrate available manganese up to a factor of 10, whereas fish and marine plants concentrate it by factors of 100 and 100 000, respectively. *Porphyra* spp. in the Irish Sea contained 13—93 mg/kg dry weight and *Fucus* spp. from British coasts contained 33—190 mg/kg dry weight (Preston et al., 1972).

All vegetation appears to concentrate manganese to some extent, the greatest degree of concentration taking place in new growth and seeds. Surface enrichment occurs through plant uptake and leaf shedding.

Aquatic and terrestrial food chains have not been fully determined for manganese. Variations reported in manganese concentrations in foods may be caused by a number of factors, such as the level and availability of manganese in the soil and water, the use of agricultural chemicals, species differences in uptake, and variations in sampling techniques and analyses.

The form in which manganese exists in animal and plant tissues is not known.

5.8 Organic manganese fuel additives

In the petrol engine, over 99% of the methylcyclopentadienyl manganese tricarbonyl (MMT) is combusted, the principal combustion product being Mn_3O_4 (Ethyl Corporation, 1974; Moran, 1975). According to available studies, less than 0.5% of MMT itself is likely to be emitted with the exhaust gas (Ethyl Corporation, 1974; Hurn et al., 1974). The emitted MMT is rapidly decomposed photochemically and has an atmospheric half-time of only a few minutes, at the most (Ter Haar et al., 1975). The photolytic decomposition products of MMT are not well known. Nearly all the manganese in this compound is converted by photochemical decomposition to a mixture of solid manganese oxides and carbonates; manganese carbonyl compounds do not appear to be formed (Ter Haar et al., 1975).

6. METABOLISM OF MANGANESE

6.1 Absorption

The main routes of absorption of manganese are the respiratory and gastrointestinal tracts. Absorption through the skin is not considered to occur to any great extent (Rodier, 1955).

6.1.1 Absorption by inhalation

Little is known about the absorption of manganese through the respiratory system. The absorption of some metals and metallic compounds was considered by the Task Group on Metal Accumulation (1973) and certain of the basic principles outlined in that group's report can be applied to inhaled metals in general. Particles small enough to reach the alveolar lining of the lung (less than a few tenths of a micrometre in diameter) are eventually absorbed into the blood. Mucociliary clearance, which differs with each individual, affects the degree of particle deposition in the lung. Furthermore, in studies by Hubutija (1972), it was shown that deposition of inhaled manganese oxide dust depended on the electrical charge carried, up to 33% more positively charged dust being deposited than negatively charged dust. As a certain percentage of inhaled manganese particles cleared by mucociliary action may be swallowed (Mena et al., 1969), absorption from the gastrointestinal tract should also be considered (Mouri, 1973).

6.1.2 Absorption from the gastrointestinal tract

Not much is known about the mechanisms of absorption of manganese from the gastrointestinal tract. From *in vitro* studies using the everted sac method, it would seem that manganese may be actively transported across the duodenal and ileal segments of the small intestine (Cikrt & Vostal, 1969). Results of studies in man and the rat on the interrelationship between manganese and iron absorption have indicated that intestinal absorption of manganese takes place by diffusion in iron-overload states and by active transport in the duodenum and jejunum in iron-deficiency states (Thomson et al., 1971).

Few quantitative data are available concerning absorption from the gastrointestinal tract in man. Mena et al. (1969) studied gastrointestinal absorption in 11 healthy, human subjects, each of whom received 100 μc (3.7 MBq) of radioactive manganese dichloride (^{54}MnCl$_2$) using 200 μg of manganese dichloride (^{55}MnCl$_2$) as a carrier. About $3 \pm 0.5\%$ of the amount administered was found to be absorbed. There were individual variations showing a five-fold difference between the lowest and highest values of absorption. The reported rate of absorption did not take into account reabsorption into the enterohepatic circulation, but the authors considered this underestimation to be small.

The rate of absorption may be influenced by such factors as dietary levels of manganese and iron, the type of the manganese compound, iron deficiency, and age. Thus, in the study just described, Mena et al. found an absorption of $7.5 \pm 2.0\%$ in 13 patients with iron-deficiency anaemia. They also found that, in 6 miners with high tissue levels of manganese, an increase in the rate of excretion of manganese was accompanied by an increase in iron excretion. This interrelationship may further aggravate a pre-existing anaemia, thus increasing the rate of manganese absorption and may be a relevant factor in occupational exposure to manganese. Similarly, Thomson et al. (1971), using duodenal perfusion with a manganese dichloride solution containing a manganese concentration of 0.5 μg/ml, noted an increased rate of absorption in iron-deficient patients that could be inhibited by adding iron to the solution.

Figures for gastrointestinal absorption in infants and young children are not available.

Most studies on animals have indicated a gastrointestinal absorption of less than 4%. Suzuki (1974) reported an intestinal absorption of only 0.5—2.0% in mice fed dietary levels [a] of manganese dichloride of 20—2000 mg/kg.

However, when rats were given 0.1 mg of radioactive manganese orally, 3—4% of the dose was absorbed (Greenberg et al., 1943). Similar results were obtained by Pollack et al. (1965), who reported an absorption of 2.5—3.5% in rats given an oral dose of radioactive manganese dichloride ($^{54}MnCl_2$). Thus, absorption data for the adult rat agree with the figure obtained for the absorption of manganese dichloride in man. However, Mena (1974) reported that intestinal absorption in the young rat was of the order of 70% compared with 1—2% in the adult rat.

In a study by Abrams et al. (1976), rats were given dietary levels of manganese ranging from 4 to 2000 mg/kg for about 2 weeks, followed by a single oral dose of radioactive manganese (^{54}Mn). Absorption of ^{54}Mn was significantly lower in rats receiving high dietary levels (1000—2000 mg/kg) than in animals receiving the lowest level (4 mg/kg).

Ethanol given to fasting rats in doses of 4 g/kg body weight increased absorption of manganese from the gastrointestinal tract and resulted in a two-fold increase in uptake of manganese in the liver. Furthermore, *in vitro* experiments indicated a four-fold increase in the transmural migration of manganese (Schafer et al., 1974). It has long been known that calcium in the diet can reduce the amount of manganese absorbed by poultry, probably by reducing the amount of manganese available for absorption (Wilgus & Patton, 1939).

However, recent studies suggest that calcium may, under certain circumstances, enhance gastrointestinal absorption of manganese. Lassiter et al. (1970) noted a higher rate of absorption in rats fed a dietary level of calcium of 6 g/kg for 21 days before oral dosing with ^{54}Mn, compared with rats receiving a level of only 1 g/kg. In studies on sheep, the same authors found that phosphoric acid, mixed

---

[a] The approximate relation between concentration in diet in mg/kg (ppm) and mg per kg body weight per day is given for a number of animal species in Nelson (1954).

into the ground hay at a concentration of 15 g/kg, decreased gastro-intestinal absorption of the stable manganese in the hay.

In rats, the enterohepatic circulation appears to be of importance. Intraduodenal administration of manganese that had been excreted into the bile resulted in about 35% absorption, whereas only 15% of an equivalent dose of manganese dichloride administered intra-duodenally was absorbed (Cikrt, 1973). This indicates that manganese present in bile is in a form that is more easily absorbed than manganese dichloride.

6.2 Distribution

6.2.1 Distribution in the human body

Manganese is an essential element for man and animals and thus occurs in the cells of all living organisms. Concentrations of manganese present in individual tissues, particularly in the blood, remain constant, in spite of some rapid phases in transport, indicating that such amounts may be considered characteristic for these particular organs irrespective of the animal species (Cotzias, 1958).

The total manganese body burden of a standard man of 70 kg has been estimated to be about 10—20 mg (Underwood, 1971; WHO Working Group, 1973; Kitamura et al., 1974). Thus, tissue concentrations will frequently be below the μg/kg level. In general, higher manganese concentrations can be expected in tissues with a high mitochondria content (Maynard & Cotzias, 1955; Thiers & Vallee,

Table 6. Manganese in human tissues (mg/kg wet weight)

Tissue	Kehoe et al. (1940) (emission spectroscopy)	Tipton & Cook (1963) [a] (emission spectroscopy)	Kitamura (1974) (atomic absorption)
aorta	—	0.11	—
brain	0.30	0.27	0.25
fat	—	—	0.07
heart	0.32	0.22	0.19
intestine	0.35	—	—
kidney	0.60	0.90	0.58
liver	2.05	1.30	1.20
lung	0.22	0.19	0.21
muscle	—	0.06	0.08
ovary	—	0.16	0.19
pancreas	—	1.18	0.74
spleen	—	0.13	0.08
testis	—	0.13	0.20
trachea	—	0.19	0.22
rib	—	—	0.06

[a] Values calculated using the given ash percentage wet weight and the median value of manganese in tissue ash.

1957), with the exception of the brain which contains only low concentrations (Maynard & Cotzias, 1955). There also appears to be a tendency towards higher concentrations in pigmented tissues such as dark hair or pigmented skin (van Koetsveld, 1958; Cotzias et al., 1964).

Table 6 gives the results of 3 studies on the manganese contents of various tissues in people without any known occupational or other additional exposure to manganese. Two are studies on adults from the USA (Kehoe et al., 1940; Tipton & Cook, 1963). In a study by Kitamura (1974) performed on 15 Japanese males and 15 females who had died in accidents, the highest concentrations of manganese were found in the liver, pancreas, kidney, and intestines. Comparatively high concentrations were also found in the suprarenal glands.

From birth to 6 weeks, infants have relatively higher tissue concentrations of manganese than older children, especially in tissues normally associated with low manganese levels. However, after about 6 weeks of age, no accumulation of manganese appears to take place with increasing age (Schroeder et al., 1966). This is in agreement with the study of Dobrynina & Davidjan (1969), who reported that manganese did not accumulate with age, and that the manganese content of the lung actually decreased with increasing age. Anke & Schneider (1974) also found a statistically significant decrease in the kidney content of manganese beginning at about 60 years of age; they reported a slightly higher mean concentration in females (4.4 mg/kg) than in males (3.8 mg/kg). With respect to manganese concentrations in the liver, Widdowson et al. (1972) reported that there was no consistent change with age in 30 fetuses from 20 weeks' gestation to full-term, but that, generally, manganese con-

Table 7. Concentrations of manganese in the whole blood of people without occupational exposure to manganese

Number of subjects	Mean (μg/100 ml)	Range (μg/100 ml)	Method	Reference
14	0.844	n.r. [a]	neutron activation	Cotzias et al. (1966)
19	n.r.	0.86—1.45	neutron activation	Cotzias & Papavasiliou (1962)
7	1.16	0.90—1.45	neutron activation	Papavasiliou & Cotzias (1961)
18	2.4	n.r.	neutron activation	Bowen (1956)
232	3.47 [b]	n.r.	spectrographic	Horiuchi et al. (1967)
47	4.0	n.r.	spectrographic	Butt et al. (1964)
12	4.6	2.2—7.9	spectrographic	Cholāk & Hubbard (1960)
13	7.6	4.0—15.0	colorimetric	Barbořik & Sehnalová (1967)
30	12.0	n.r.	spectrographic	Kehoe et al. (1940)

[a] n.r. = not reported.
[b] median.

Table 8. Concentrations of manganese in the plasma and serum of people without occupational exposure.

Number of subjects	Mean (μg/100 ml)	Range (μg/100 ml)	Method	Reference
12 (S)[a]	n.r.[c]	0.036—0.090	colorimetric	Fernandez et al. (1963)
14 (P)[b]	0.059	n.r.	neutron activation	Cotzias et al. (1966)
25 (S) (F)[d]	0.055	0.038—0.104	neutron activation	Versieck et al. (1974a)
25 (S) (M)[e]	0.059	0.045—0.101	neutron activation	Versieck et al. (1974a)
19 (P)	n.r.	0.183—0.310	neutron activation	Cotzias & Papavasiliou (1962)
7 (P)	0.269	0.210—0.302	neutron activation	Papavasiliou & Cotzias (1961)
16 (S)	0.250	0.205—0.297	neutron activation	Papavasiliou & Cotzias (1961)
7 (P)	0.18[f]	n.r.	neutron activation	Hagenfeldt et al. (1973)
—	0.32[g]	n.r.	neutron activation	Hagenfeldt et al. (1973)
15 (P)	0.43	n.r.	neutron activation	Olehy et al. (1966)
90 (S) (F)	1.05	n.r.	spectrographic	Žernakova (1967)
60 (S) (M)	0.96	n.r.	spectrographic	Žernakova (1967)
48 (S)	1.3	n.r.	spectrographic	Butt et al. (1964)
30 (S)[h]	1.3	0.9—1.9	neutron activation	Kanabrocki et al. (1964)
40 (S)	2.4	1.2—3.8	atomic absorption	Mahoney et al. (1969)

[a] (S) = serum, [b] (P) = plasma, [c] n.r. = not reported. [d] (F) = female, [e] (M) = male. [f] sampled at days 16—18 of a menstrual cycle. [g] sampled at days 6—8 of a menstrual cycle. [h] non-dialysable serum.

centrations in full-term livers were 7—9% higher than concentrations in adult livers. Studies by Schroeder et al. (1966) and Widdowson et al. (1972) confirmed that human placental transfer of manganese takes place.

Some reports on the manganese contents of whole blood, plasma, and serum have been summarized in Tables 7 and 8. All studies were on subjects without any occupational exposure to manganese. The concentrations of manganese are low in blood and still lower in plasma and serum, thus increasing the vulnerability of sampling and analytical procedures to the possibilities of contamination (section 2.2.1). This may partly explain the wide range of manganese concentrations found in the literature. Support for this theory comes from Cotzias et al. (1966), who considered that systematic contamination was responsible for the fact that previous plasma levels obtained by this group (Papavasiliou & Cotzias, 1961; Cotzias & Papavasiliou, 1962) were 4 times higher than those obtained in the 1966 study (Table 8). A low order of magnitude of manganese levels in plasma and serum was reported by Fernandez et al. (1963) and more recently by Versieck et al. (1973b, 1974a).

The concentrations of manganese in blood and serum appear to be fairly stable over long periods of time (Cotzias et al., 1966; Mahoney et al., 1969). A slight seasonal variation in blood manganese concentration, has been reported, the levels being somewhat lower during the summer and autumn months (Horiuchi et al., 1967).

45

In this study, manganese concentrations in blood did not differ between age groups. Diurnal variations were reported by Šabadas (1969), the concentrations in blood being higher during the day than during the night. There do not appear to be any differences in the concentrations of manganese in the blood of men and women (Horiuchi et al., 1967; Zernakova, 1967; Mahoney et al., 1969; Versieck et al., 1974a).

Hegde et al. (1961) claimed that manganese concentrations in serum increased following myocardial infarction, but, in more recent studies, Versieck et al. (1975) were unable to detect such a relationship. However, during the active phase of hepatitis, serum concentrations of manganese were invariably elevated (Versieck et al., 1974b).

The mean concentration of manganese in the urine of unexposed people has been reported to be in the range of 3—21 μg/litre (Kehoe et al., 1940; Cholak & Hubbard, 1960; Horiuchi et al., 1967; Tichý et al., 1971; McLeod & Robinson, 1972b).

6.2.2 Distribution in the animal body

Average levels of manganese in unexposed rabbit tissues were reported to be: 2.1 mg/kg wet weight in the liver; 2.4 mg/kg in the pituitary; 1.6 mg/kg in the pancreas; and 1.2 mg/kg in the kidney. The brain has a relatively low average content of 0.4 mg/kg wet weight (Fore & Morton, 1952). The lowest levels occur in bone marrow (0.04 mg/kg wet weight), blood (0.03 mg/kg wet weight) and lung (0.01 mg/kg wet weight) (Cotzias, 1958). According to Suzuki (1974), when aqueous solutions of manganese dichloride at concentrations ranging from 20 to 2000 mg/litre were given to mice, concentrations below 500 mg/litre did not result in accumulation in the organs. However, there was distinct accumulation at doses exceeding 1000 mg/litre. When mice were exposed through inhalation to manganese dioxide concentrations of 5.6 and 8.9 mg/m^3 (particle size 3 μm) every 2 h, for 8 and 15 days, respectively, the highest concentrations of manganese were found in the kidney (10.8 and 8.4 mg/kg dry weight), liver (9.0 and 7.1 mg/kg), pancreas (8.4 and 8.2 mg/kg), and brain (5.9 mg/kg). Because of the route of administration, even higher concentrations were observed in the lungs, trachea, and gastrointestinal tract (Mouri, 1973).

In a study in which monkeys were given a subcutaneous injection or a suspension of manganese dioxide once a week, for 9 weeks, the manganese concentration increased markedly in the tissues of the endo- and exocrine glands and of the cerebral basal ganglion, the accumulation rate being proportional to the dose administered (Suzuki et al., 1975).

After intraperitoneal administration of radioactive manganese to rats, the highest concentrations were found in the suprarenal, pituitary, liver, and kidney tissues (Dastur et al., 1969). The uptake by glandular structures was also high in monkeys after intraperitoneal injection of radioactive manganese (Dastur et al., 1971).

Several experimental studies have shown that manganese penetrates the placental barrier of various species (Koshida et al., 1963; Järvinen & Ahlström, 1975; Miller et al., 1975). It has been reported that manganese is more uniformly distributed in the fetal tissues of the mouse than in adult tissues. The main differences were seen in the concentrations in kidney and liver tissues, which were lower in the fetuses than in the adults. At a later embryonic stage, manganese accumulation in the bone took place parallel with the process of ossification (Koshida et al., 1963; 1965). Miller et al. (1975) showed that, in contrast to adult animals and man, neonatal mice did not excrete manganese during the first 17—18 days of life, despite vigorous absorption of the radioactive metal (^{54}Mn) with accumulation in both mitochondria and tissues, notably in the brain. This suggested an initial accumulation of the essential micronutrient, supplied in trace amounts in mouse milk (54 μg/litre) by mothers consuming much higher dietary concentrations (50 mg/kg). In subsequent experiments, the same authors noted an absence of manganese excretion during the first 18 days of life in neonatal rats and kittens. Moreover, when lactating mothers were fed diets containing concentrations of manganese ranging from 40 to 40 000 mg/litre, the lactation barrier appeared to give adequate protection to the young. However, when the dietary level exceeded 280 mg/litre, the newborn initiated excretion before the 16th day of life. The neonates showed a greater accumulation than their mothers, whereas the increase in liver concentrations was proportional to concentrations found in the mother's liver. The findings suggest that the neonatal brain may be at higher risk of reaching abnormal concentrations than other tissues. In the Syrian hamster, manganese was found in embryonic tissues 24 h after intravenously injecting the mother with radioactive manganese at 1.36 mg/kg body weight (Hanlon et al., 1975).

After a single intragastric dose of 0.5 or 2.5 mg of methylcyclopentadienyl manganese tricarbonyl (MMT) labelled with ^{54}Mn, rat tissue and organs showed a distribution characteristic of inorganic manganese, i.e., the highest concentrations were found in the liver, kidney, and pancreas. However, high concentrations were also found in the lungs and to lesser degree in abdominal fat (Moore et al., 1974).

6.2.3 Transport mechanisms

Absorbed manganese is concentrated in the liver and it has been suggested that it forms complexes with bile components (Tichý &

Cikrt, 1972). It has also been suggested that manganese is transported directly into the bile (Klaassen, 1974). At one time, it was thought to be transported in the plasma in its trivalent form by a β_1-globulin other than transferrin, called transmanganin (Cotzias, 1962), but the results of later *in vitro* studies on the serum of cows and human serum and *in vivo* studies on rabbit and rat blood have refuted this theory (Panić, 1967, Hancock et al., 1973). At present, it is largely accepted that manganese and iron are both transported by the transferrin in the plasma (Panić, 1967; Mena et al., 1974). It has already been pointed out that mitochondria have been shown to contain a non-dialysable fraction of manganese (Fore & Morton, 1952). Studies using radioactive manganese (^{56}Mn) indicated that newly deposited ^{56}Mn was easily removed from mitochondria whereas older, stable manganese was not, suggesting different types of bonding with mitochondria (Cotzias, 1958).

6.3 Biological Indicators of Manganese Exposure

Estimation of manganese exposure in man by examination of biological fluids or tissues has not proved to be a reliable index. Analyses of blood or urine samples from persons with signs and symptoms of manganese poisoning do not usually reveal high levels of manganese. However, a rough correlation between urine levels and average air concentrations seems to exist (Tanaka & Lieben, 1969). There is also some evidence that manganese may concentrate in hair following exposure to high concentrations (Rosenstock et al., 1971). Because of the short biological half-time, manganese levels in tissues and organs can only be related to recent exposure. At present, no specific diagnostic biological materials are known that could be used to monitor manganese exposure in epidemiological studies or for clinical diagnosis.

6.4 Elimination

It has already been pointed out that tissue concentrations of manganese are remarkably constant without any tendency to accumulate with age, after the first few weeks of life. From earlier studies, it was considered that variable excretion rather than variable absorption played an important role in manganese homeostasis (Britton & Cotzias, 1966). However some later results, from studies using oral dosing in rats, have indicated that variable absorption is also an important factor (Abrams et al., 1976).

The manganese absorbed in the body, whatever the route of absorption, is eliminated almost exclusively in the faeces. At ordinary exposure levels, manganese is mainly excreted into the bile (Papavasiliou et al., 1966). Quantitative data concerning excretion in man are not available. After intravenous injection of 0.6 μg of manganese dichloride ($MnCl_2.4H_2O$) in rats, 12 % of the injected dose was excreted into the bile within 3 h (Tichý et al., 1973), and 27% within 24 h (Cikrt, 1972). Intraperitoneal administration of 0.01 mg of manganese to rats resulted in the biliary excretion of 26% of the dose within 48 h; at a dose of 0.1 mg, the fraction appearing in the bile was 37% (Greenberg et al., 1943). Manganese excreted with the bile flow into the intestine is partly reabsorbed (section 6.1.2). In the rat, there is some evidence of the excretion of manganese through the intestinal wall into the duodenum, jejunum and, to a lesser extent, the terminal ileum (Bertinchamps et al., 1966; Cikrt, 1972).

In dogs, some manganese is also excreted with the pancreatic juice (Burnett et al., 1952). It has been shown that, while excretion by the biliary route predominates under normal conditions, excretion by the auxiliary gastrointestinal routes may increase in significance in the presence of biliary obstruction or with overloading with manganese (Bertinchamps et al., 1966; Papavasiliou et al., 1966). Results of human studies have shown that only a small amount of manganese (about 0.1—1.3% of the daily intake) is excreted through the kidneys into the urine (Maynard & Fink, 1956; Tipton et al., 1969; McLeod & Robinson, 1972b). Urinary excretion is not increased by biliary obstruction or by overloading (Papavasiliou et al., 1966), but in rats it was increased many times by the administration of ethylenediamine tetraacetic acid (EDTA), urine becoming the predominant excretory route for 24 h, after which time faecal elimination was resumed (Kosai & Boyle, 1956; Maynard & Fink, 1956).

Following intravenous administration of radiolabelled methylcyclopentadienyl manganese tricarbonyl (MMT) and manganese dichloride to rats, retention of [54]Mn was similar, but the route of excretion was different. After intravenous administration of radioactive manganese chloride, only a trace was detected in urine, [54]Mn being excreted in the faeces. However, both oral and intravenous administration of MMT resulted in excretion in both faeces and urine. With oral administration of MMT, the urine/faeces ratio of manganese varied from 0.68 to 0.25. No MMT was detectable in the faeces, indicating a biotransformation of MMT. *In vitro* experiments showed that MMT was metabolized in the liver, lung and kidney and to a small extent in the brain. Biotransformation of MMT by kidney homogenate may explain the high concentrations of manganese found in the urine of the rats (Moore et al., 1974).

Insufficient data are available on dermal losses of manganese. An average excretion in human sweat was given as 60 μg/litre by

Mitchell & Hamilton (1949). Although sweat volumes are known to vary over a wide range (International Commission on Radiological Protection, 1975), the average daily excretion of manganese with sweat is probably in the range of 30—120 μg, assuming a sweat volume of 0.5—2 litre/day. This is in good agreement with a study by Conzolazio et al. (1964), who found a daily mean excretion of about 100 μg (corresponding to 2.3% of the total daily intake of manganese) in 3 men exposed to 37.8°C for 7.5-h periods. The loss of manganese with human hair and nails has been estimated to about 2 μg/day (International Commission on Radiological Protection, 1975), although allowance must be made for the considerable variation of manganese concentrations in hair (Cotzias et al., 1964). Small quantities of manganese are also transferred through the placenta (section 6) and the excretion with breast milk is about 10—20 μg/day (McLeod & Robinson, 1972a).

6.5 Biological Half-times

A few studies have been performed on animals and man in order to assess the biological half-times of inorganic manganese. However, half-times for organic forms of manganese have only been studied in animals. When considering biological half-times of trace elements in living organisms, the following factors should be taken into account:

(a) type of exposure (oral or parenteral, single or multiple);
(b) type of metabolic model (single compartment or multi-compartment);
(c) inter- and intraspecies variations; and
(d) the rates of phases of excretion (rapid and slow components).

6.5.1 Man

Mahoney & Small (1968) used single injections of radiolabelled manganese dichloride on 6 human volunteers to study the biological half-time of manganese in man. They found 2 phases in the elimination of manganese from the body, one of which was slow and the other fast. The mean biological half-time for the 3 "normal" adults was about 4 days for the fast phase and 39 days for the slow phase. About 60—65% of elimination occurred during the slow phase, although in one subject 90% was eliminated during this phase. In subjects with a high oral intake of manganese, elimination from the body took place at an increased rate. Iron reserves may also influence the retention of manganese dichloride.

Cotzias et al. (1968) studied the tissue clearance of manganese in 19 healthy "normal" volunteers after a single injection of [54]Mn.

Clearance half-time of [54]Mn was 37.5 days for the whole body, 25 days for the liver, 57 days for the thigh, and 54 days for the head. The clearance half-time from blood and plasma was less than 1.5 minutes.

A clearance half-time for the whole body of 37 days was reported by Mena et al. (1969) in healthy subjects, compared with a half-time of 23 days in iron-deficient, anaemic patients. They also reported a half-time of 34 days in miners suffering from chronic manganese intoxication, while healthy miners exhibited a fast turnover of only 15 days. This finding may have a bearing on the question of individual susceptibility to manganese intoxication.

6.5.2 Animals

Britton & Cotzias (1966) reported a two-component whole-body clearance of manganese in mice comparable with that reported by Mahoney & Small (1968) in man. The half-time of the slow component was given as 50 days and that of the fast component as 10 days. The fraction eliminated with the slow component was lower in mice than in man, i.e., approximately 35%. However, on a low-manganese diet, it increased to about 95% and the half-time for this component decreased from 50 to about 35 days. With a high-manganese diet, the half-time for the fast component decreased from 10 to 2 days.

The effect of dietary manganese levels on the biological half-time of manganese was studied in mice by Suzuki (1974). The animals received an aqueous solution of manganese dichloride in concentrations ranging from 20 to 2000 mg/litre for 26—30 days, after which radioactive manganese was administered. The whole body clearance half-time was about 6 days in the 20 mg/litre group, about 3 days in the 100 mg/litre group, and 1—1.5 days in the group that had received 2000 mg/litre, i.e., the heavier the preloading, the more rapid was elimination from the whole body. The half-time in the brain was longer than that in the whole body.

Chemobiokinetic studies on [54]Mn in monkeys disclosed a half-time of 95 days for whole-body elimination. Brain levels did not decrease significantly over the experimental period of 278 days, suggesting that the clearance half-time in the brain was longer than that for the whole body (Dastur et al., 1871).

Adrenal glucocorticoids accelerated the total body clearance of manganese in mice (Hughes & Cotzias, 1960), and later studies using ACTH stimulation suggested the existence of an adrenal regulatory mechanism for the metabolism of manganese (Hughes et al., 1966).

The whole-body retention curve for methylcyclopentadienyl manganese tricarbonyl (MMT) in rats was similar to that for manganese dichloride (section 6.4). The authors considered that this was due to the rapid metabolism of MMT and that the retention curve only reflected the metabolism of the labelled manganese (^{54}Mn) (Moore et al., 1974).

7. MANGANESE DEFICIENCY

7.1 Metabolic Role of Manganese

The essential role of manganese as a trace metal nutrient for mammals was discovered mainly through experimental and epidemiological studies of deficiency states in animals. Thus, manganese has been shown to be associated with the formation of connective tissue and bone, with growth, carbohydrate and lipid metabolism, the embryonic development of the inner ear, reproductive function, and, probably, brain function (Underwood, 1971; NAS/NRC, 1973).

The biochemical background of the metabolic defects that have been observed is poorly understood, though a few specific biochemical properties of manganese have been discovered. One is that manganese catalyses the formation of glucosamine-serine linkages in the synthesis of the mucopolysaccharides of cartilage; another, that the mitochondrial enzyme pyruvate carboxylase (EC 6.4.1.1) is a manganese metalloenzyme, thus, manganese is linked with carbohydrate metabolism. It has also been discovered that the digestive enzymes prolidase (EC 3.4.13.9) and succinic dehydrogenase (EC 1.3.99.1) are manganese-dependent and that, *in vitro*, manganese can substitute for other metals, especially magnesium, in various biological reactions (Underwood, 1971; NAS/NRC, 1973). Lindberg & Ernster (1954) performed *in vitro* experiments on rat liver mitochondria that demonstrated that manganese was required as a cofactor in oxidative phosphorylation. Manganese deficiency in mice was reported by Hurley (1968) to be associated with a decreased oxygen uptake by the liver mitochondria. However, the relationship between these findings and defects due to manganese deficiency remains obscure.

7.2 Manganese Deficiency and Requirements in Man

No definite syndrome of manganese deficiency has been described in man. However, in a human subject with experimentally induced vitamin K deficiency, a sequence of signs was attributed to the accidental omission of manganese from the diet during 1 week; the daily intake during 16 subsequent weeks was retrospectively calculated to have been about 0.35 mg. The subject was unable to elevate the depressed clotting proteins in response to vitamin K and this finding was experimentally reproduced in the chick. Moreover, marked hypocholesterolaemia, retarded growth of hair and nails, mild dermatitis, pigment changes in hair and beard, and moderate weight loss were present (Doisy, 1973).

It has been difficult to estimate the minimum physiological requirements of manganese for man. On the basis of existing data on the daily manganese intake and manganese balance in man, a WHO Expert Committee concluded that an intake of 2—3 mg/day was adequate for adults (WHO, 1973). This is compatible with the figures quoted in section 4.5 and also agrees with the estimate of 2.7 mg made by De (1949) on the basis of balance studies on male subjects. A negative manganese balance was recorded by De in subjects with a mean manganese intake of 0.71 mg per day.

Engel et al. (1967) measured the daily intake of manganese in 6 to 10-year-old girls and estimated that 1 mg per day was needed to maintain the balance. Taking into account growth needs, integumental losses, and a reasonable safety margin, they suggested a required daily intake of 1.25 mg (0.045 mg/kg body weight). This also agrees well with the data presented in section 4.5, which show that healthy children within the age ranges of 9—13 and 3—5 years had daily intakes of 0.06 and 0.08 mg/kg body weight, respectively (Schlage & Wortberg, 1972). A positive manganese balance was observed in girls, aged 7—9 years, with a daily intake of 2.1—2.4 mg of manganese (Price et al., 1970). Breastfed infants may have a daily intake as low as 0.002—0.003 mg/kg (Widdowson, 1969; McLeod & Robinson, 1972b) and may exhibit a distinctly negative manganese balance durin the first week of life. Widdowson (1969) reported that the amount excreted in the faeces during the first week amounted to 3—5 times the amount ingested daily with breast milk, indicating the excretion of manganese from tissue reserves that had accumulated during fetal life (section 6.2.1). This is in agreement with the observation of Schroeder et al. (1966) that tissue levels decreased during the first 45 days of life.

An association has been suggested between manganese deficiency and lupus erythematosus found in patients following treatment with hydralazine. This is based on the fact that administration of manganese(II) salts improved the condition of such patients and of patients suffering from the spontaneous variety of disseminated lupus erythematosus (Comens, 1956).

7.3 Manganese Deficiency in Animals

Defects due to manganese deficiency have been shown experimentally in a variety of laboratory animals. The best-documented manifestations are those associated with skeletal abnormalities and impaired growth. Abnormally fragile bones that are shorter than normal, and bowed forelegs resulting from these changes have frequently been reported in mice, rats, and rabbits (Amdur et al., 1945; Ellis et al., 1947; Plumlee et al., 1956). Perosis with deformity of bones and dislocation of the achilles tendon ("slipped tendon") has long been known in young chickens (Wilgus & Patton, 1939). Early changes observed in chick embryos include nutritional chondrodystrophy, retarded growth, and shortening of the lower mandible (Lyons & Insko, 1937). In his review of manganese deficiency, Underwood (1971) quotes crooked and shortened legs and enlarged hock joints in pigs, leg deformities with "overknuckling" in cattle, joint pains in sheep, and tarsal joint excrescences in goats.

Ataxia with loss of equilibrium and altered postural reactions to stimuli, without histological changes in the brain tissue have been reported in rats (Hill et al., 1950). It is likely that ataxia is due to the impaired development of the inner ear, where the otoliths appear to be absent or defective in several species of laboratory animals (NAS/NRC, 1973).

Manganese deficiency can induce congenital malformations, still-births, and neonatal deaths in rats and guineapigs and seminal tubular degeneration and aspermia have been observed in rats and rabbits (NAS/NRC, 1973). In manganese-deficient female rats, estrous cycles may be absent or irregular, and the rats may be sterile; in severe deficiency states, the animals will not mate (Underwood, 1971).

A decreased tolerance to orally administered glucose and an impaired peripheral use of parenterally administered glucose have been reported in guineapigs (Everson & Shrader, 1968). Newborn offspring of manganese-deficient guineapigs displayed aplasia or distinct hypoplasia of all cellular components of the pancreas; islets contained a reduced number of beta cells, which were also less intensely granulated (Shrader & Everson, 1968).

There have been some reports relating manganese to lipid metabolism. Interaction between choline and manganese has long been recognized (Underwood, 1971). Pigs, fed manganese-deficient food, showed a statistically significant increase in fat deposits, measured as back-fat thickness (Plumlee et al., 1956). Curran (1954) reported that manganese stimulated the synthesis of fatty acids and cholesterol in the rat liver. Although the biological implications of these findings are not clear, it has been suggested that a decrease in the synthesis of cholesterol and its precursors due to manganese deficiency might limit the synthesis of sex hormones. This could explain the sterility of manganese-deficient animals (Doisy, 1972).

A reduction in liver arginase (EC 3.5.3.1) activity in manganese-deficient rats and rabbits has been reported, but the importance of this remains to be assessed (Underwood, 1971).

Table 9. The acute toxicity of various forms of manganese

Compound	Animal	Administration route	LD$_{50}$ (mg/kg)	Reference
manganese dioxide	mouse	subcutaneous	550 [b]	Date (1960)
manganese chloride	mouse	oral	275—450	Sigan & Vitvickaja (1971)
	rat	oral	250—275	Hazaradze (1961)
	guineapig	oral	400—810	Hazaradze (1961)
manganese sulfate	mouse	intraperitoneal	64	Yamamoto & Suzuki, (1969)
	mouse	subcutaneous	146 [b]	Date (1960)
	mouse	oral	305 [b]	Date (1960)
manganese nitrate	mouse	intraperitoneal	56	Yamamoto & Suzuki (1969)
potassium permanganate	mouse	subcutaneous	500 [b]	Date (1960)
	mouse	oral	750	Sigan & Vitvickaja (1971)
	rat	oral	750	Sigan & Vitvickaja (1971)
	guineapig	oral	810	Sigan & Vitvickaja (1971)
DAP-Mn cake [a]	mouse	oral	> 8000	Suzuki et al. (1972)
	mouse	intraperitoneal	> 1200	Suzuki et al. (1972)
DAP-Mn dust [a]	mouse, male	oral	2790	Suzuki et al. (1972)
	mouse, female	oral	2570	Suzuki et al. (1972)
	mouse, male	intraperitoneal	378	Suzuki et al. (1972)
	mouse, female	intraperitoneal	352	Suzuki et al. (1972)

[a] DAP refers to a process for the removal of sulfur dioxide from flue gas. DAP-Mn cake: manganese oxides used in the desulfurization process. DAP-Mn dust: exhaust gas and dust from the desulfurization process in a plant.
[b] Lethal dose (LD$_{100}$).

8. EXPERIMENTAL ANIMAL STUDIES ON THE EFFECTS OF MANGANESE

8.1 Median Lethal Dose

The toxicity of manganese varies according to the chemical form administered. Divalent manganese has been shown to be 2.5—3 times more toxic than the trivalent form. The median lethal doses (LD$_{50}$) of various forms of manganese are listed in Table 9.

8.2 Effects on Specific Organs and Systems

8.2.1 Central nervous system

Attempts to induce brain damage characteristic of manganese intoxication by feeding manganese compounds to experimental animals have not been completely successful (van Bogaert & Dallemagne, 1946). This may partly be due to the low absorption of orally administered manganese. Šigan & Vitvickaja (1971) showed that potassium permanganate altered conditioned reflex activity in rats, when administered orally at 10 mg/kg body weight per day for 9 months, and to a lesser extent when administered at a dose of 1 mg/kg.

Exposure of a monkey to manganese dioxide aerosol at concentrations of 0.6—3.0 mg/m³ for 95 1-h periods during 4 months, initially produced alternating periods of sudden movement and torpor, nervousness, severe tremor, flexion-extension movements of upper limbs, yawning, and cyanosis. Sequelae after 5 months included gross tremors, uncertain gait, and paresis. Histological examination of the brain revealed atrophy of the cerebellar cortex, whereas the putamen, caudate, and pallidum did not exhibit any clear changes (van Bogaert & Dallemagne, 1946). Intraperitoneal administration of manganese dichloride to 4 monkeys on alternate days for up to 18 months, starting with a dose of 5 mg and increasing the dose to between 15 and 25 mg, resulted in the characteristic lesions of the basal ganglia associated with choreiform movements and muscular rigidity, tremors, and limb contractions. Histologically, lesions with shrunken and pyknotic cells were detected in the putamen, caudate, and the pallidum. Demyelinized fibres were found in these areas. A slight chromatolysis was found in the cortex of *Macacus rhesus* (Mella, 1924). In studies on monkeys by Osipova et al. (1968), a clinical picture suggesting extrapyramidal dysfunction was obtained in animals given manganese dichloride subarachnoideally at 1—2 mg/kg body weight, in 3—8 doses. Similar effects were produced in 5 monkeys given repeated intramuscular administrations of manganese dioxide at several injection sites (Pentschew et al., 1963). In this study, the doses administrated (2000 and 3500 mg injected with an interval of 3 months) were reported for only one monkey, which was sacrificed 14¹/₂ months after the first injection. The histopathological findings in the monkey mainly involved the subthalamic nucleus and the medial and lateral pallidum and were characterized by proliferation of bizarre cells and by extensive loss of neurons. Diffuse alterations were reported in the cerebrum, brainstem, and cerebellum. A study on early brain lesions in rats was reported by Chandra & Srivastava (1970). Animals which had received intraperitoneal administrations of manganese dichloride (8 mg/kg body weight) were sacrificed at 30-day intervals. The first histological changes, seen at 120 days, consisted of neuronal degeneration in the cerebellar and cerebral cortex. The extent of brain lesions increased in intensity up to 180 days and was directly related to the amount of manganese present in the brain tissue.

When monkeys were given subcutaneous injections of a suspension of manganese dioxide, once a week for 9 weeks, at doses of 0.25—1.0 g, the typical extrapyramidal symptoms that appeared after

3—4 weeks were not proportional to the dose administered. However, the time of appearance of the symptoms was dose-related (Suzuki et al., 1975). Symptoms such as muscular rigidity and tremor were induced in squirrel monkeys by subcutaneous injection of a total of 400 mg of manganese dioxide divided into 2 doses, administered with a 5-week interval. These animals did not show any histological abnormalities of the brain, when sacrificed $3^1/_2$ months after the first injection, indicating that symptoms and signs as well as biochemical changes may appear before histological alterations can be found (Neff et al., 1969).

8.2.2 Respiratory system

"Manganese pneumonitis" is associated with inhalation of fine dust containing a relatively low concentration of manganese dioxide and probably other oxides of manganese, and does not lead to permanent pulmonary changes or fibrosis. Acute pneumonitis can be induced in rats by intratracheal administration of manganese dioxide dust or a solution of manganese dichloride. Characteristically, shedding of bronchial and alveolar epithelium is seen with intense mononuclear cell infiltration of the alveolar walls and alveoli. In a study on rats by Lloyd Davies & Harding (1949), intratracheal administration of manganese dioxide solutions produced intense mononuclear infiltration of the alveolar walls and alveoli, followed by granulomatous and giant cell formation. The changes disappeared within a year. Pulmonary congestion and oedema were observed after intratracheal injection of manganese dichloride. Similar intense mononuclear infiltration of the alveolar walls was produced in young rats by intratracheal administration of suspensions of various manganese compounds (particle size less than 3 μm). The higher oxides of manganese and freshly prepared solutions were more toxic (Levina & Robačevskaja, 1955).

Peribronchial and perivascular sclerosis and the appearance of collagenic threads were observed in rats after intratracheal administration of either 25 mg of ferromanganese dust containing 10 mg of manganese dioxide once a month for 4 months, or 10—30 mg of manganese dioxide in 6—10 doses over a period of $7^1/_2$ months (Levina & Robačevskaja, 1955). Similarly, peribronchial and perivascular sclerosis and inflammatory changes were seen in 15 out of 15 female rats exposed by inhalation to manganese dioxide at a mean concentration in air of 0.3 mg/m³, for 5—6 h daily, for 6 months but not in rats exposed to 0.033 mg/m³ for 5—6 daily for 7 months (Dokučaeva & Skvortsova, 1966).

Nishiyama et al. (1975) exposed groups of 3 and 2 monkeys, 20 rats, and 70 mice to air concentrations of manganese dioxide of 3 mg/m³ and/or 0.7 mg/m³ (particle size less than 1 μm), for

22 h daily, over a period of 5 months. Thorax X-rays appeared mottled in both groups of monkeys after 2—4 months of exposure. In 2 monkeys exposed to 3 mg/m³, the mottled picture appeared after 2 months, the patches growing larger and increasingly confluent at 3—5 months. The X-rays of these monkeys showed accentuation of blood vessels, indicating pulmonary congestion. The third monkey in this group and 2 control animals did not show any signs of adverse effects. The 2 monkeys exposed to a level of 0.7 mg/m³ showed abnormal findings that were less severe and appeared at a later stage, i.e., at 3—4 months. Mice showed inflammatory changes after 2 weeks at both levels of exposure. The inflammation disappeared after 2 months, at which time desquamation of bronchial epithelium was observed.

Maigretter et al. (1976) exposed mice to aerosols containing a manganese dioxide concentration of 109 mg/m³ for one or more 3-h periods, after which the animals where challenged with *Klebsiella pneumoniae* or influenza A virus. A decrease in resistance to infection was observed, even after a single exposure to the manganese aerosol. The authors discussed the possible causal association between the reduced resistance to infection and *in vitro* observations showing that manganese dioxide reduced the number and viability of alveolar macrophages (Walters et al., 1975), and also impaired the phagocytic activity of these cells (Graham et al., 1975).

Studies have been reported indicating the possibility of a synergistic effect of manganese dioxide and sulfur dioxide on the lungs of guineapigs (Rylander et al., 1971; Rylander & Bergström, 1973). The ability to clear inert particles was lower in guineapigs exposed for 4 weeks (6 h per day, 5 days per week) to manganese dioxide (5.9 mg/m³) + sulfur dioxide (14.2 mg/m³) than in control animals or animals exposed to either of the two compounds singly (Rylander et al., 1971). Exposure of guineapigs to a combination of manganese dioxide at 20 mg/m³ and sulfur dioxide at 56.8 mg/m³ resulted in a marked increase in leukocytes in the lungs, and a histological evaluation of the tracheal epithelium gave an "irritation score" of 3.0 for exposed animals compared with 1.8 for controls. Animals exposed to sulfur dioxide alone had a score of 1.9, while the score was 2.3 for those exposed to manganese dioxide only (Rylander & Bergström, 1973).

8.2.3 Liver

Rats given an intravenous dose of manganese at a concentration of 55—60 mg/kg body weight showed a marked decrease in the ability of the liver to clear bilirubin from the bile. This was associated histologically with cholestatic changes. Reduction in bile flow occurred within 4 h (Witzleben, 1969). However, no decrease in bile flow was noted in rats 24 h after an intravenous dose of manganese dichloride of 30 µg (Cikrt, 1972).

A subcutaneous dose of 170 mg/kg body weight produced hepatic necrosis in rats within 18 h (Baxter et al., 1965). The metabolism of thiamine appears to be linked with that of manganese, the storage of manganese in the liver being related to the level of thiamine in the diet (Hill & Holtkamp, 1954). Subcutaneous injection of 1.5 and 25 mg/kg body weight of manganese sulfate produced a decrease in monoamine oxidase (EC 1.4.3.4.) activity in rat liver. The reduction in activity was more pronounced after repeated injections of 5 and 25 mg/kg for 10 and 5 days, respectively (Levina & Tčekunova, 1969).

Rats exposed to manganese dioxide by inhalation for 5—6 h daily, 6 times a week for 7 months, displayed a decreased serum albumen/globulin ration, which might have been a result of liver effects (Dokučaeva & Skvorcova, 1966).

Ultrastructural alterations were found in the liver cells of rats after administration of manganese dichloride in the drinking-water at 200 mg/litre for 10 weeks. The treated animals exhibited an increased amount of rough endoplasmic reticulum, a proliferated smooth endoplasmic reticulum in the centrolobular area, prominent Golgi apparatus in the biliary area, numerous mitochondria which were sometimes polymorphous and frequently had an electron-dense matrix. The changes suggested a process of adaptation to increased exposure to manganese dichloride (Wassermann & Wassermann, 1977).

Rabbits given 3.5 mg per day of manganese dichloride intravenously, for 32 days, developed hepatic congestion, central vein thrombosis, and focal necrosis with leukocyte infiltration (Jonderko & Szczurek, 1967).

Other effects of manganese possibly associated with the liver metabolism that have been observed include an increase in cholesterol synthesis in the rat (Curran, 1954), a disturbance in lipid and carbohydrate metabolism with lipid deposition in the liver and adrenals (Roščin, 1971), and an enhancement of the coagulating activity of the blood (Cereteli & Kipiani, 1971).

8.2.4 Cardiovascular effects

As early as 1883, Kobert noted that manganese could produce a reduction in blood pressure. Antihypertensive effects in the rat and the cat were also reported by Schroeder et al. (1955) and Kostial et al. (1974). Mjačina (1972) found an increase in the activity of monoamine oxidase (EC 1.4.3.4) in the cardiac tissue of rats following subcutaneous injection of manganese dichloride at 10 mg/kg body weight.

8.2.5 Haematological effects

Rats given manganese dichloride in doses of 50—1500 mg/kg body weight showed an increase in haemoglobin and haematocrit, mean corpuscular volume, and serum chloride, phosphate, and magnesium after 4 h (Baxter et al., 1965). In anaemic lambs, manganese levels in diet of 1000—2000 mg/kg caused a retardation in haemoglobin regeneration and a decrease in serum iron concentrations (Hartman et al., 1955). At a dietary level of 2000 mg/kg, haemoglobin formation was also depressed in anaemic rabbits and baby pigs. The effect in pigs was reversed by a dietary supplement of iron of 400 mg/kg (Matrone et al., 1959). Dietary levels of manganese ranging from 50 to 125 mg/kg were found to be the minimum levels that would interfere with the formulation of haemoglobin in the baby pigs. The minimum level for lambs was 45 mg/kg (Hartman et al., 1955).

8.3 Effects on Reproduction

Disturbances in sex function and testicular changes have been noted in rats following exposure to potassium permanganate. Animals exposed orally or by inhalation to doses of 50 mg/kg body weight for various periods of time exhibited changes in spermatogenesis. Embryogenesis was also adversely affected (Mandzgaladze, 1966b, 1967).

In studies on rabbits, intravenous administration of manganese dichloride at 3.5 mg/kg body weight was reported to produce histochemically detectable alterations in the testes, confirmed by decreases in NADH diaphorase, succinic dehydrogenase (1.3.99.1), and glucose-6-phosphate-dehydrogenase (1.1.1.49) activities. These changes affected germinal activity (Iman & Chandra, 1975).

8.4 Carcinogenicity

Few experimental studies have been conducted on the carcinogenicity of manganese and its compounds. In one recent study, manganese(II) sulfate was administered intraperitoneally to mice over a period of 30 weeks. The highest dose of 10 mg/kg body weight (15 injections) produced a statistically significant ($p < 0.05$) increase in the incidence of lung tumours in treated animals compared with controls (Stoner et al., 1976). There are no other available studies indicating that inorganic manganese compounds are carcinogenic.

A commercial fungicide containing manganese ethylene-bis-dithiocarbamate was evaluated for carcinogenicity by the International Agency for Research on Cancer (IARC Working Group, 1976). Mice of 4 different strains were given 6 weekly doses of the compound at the rate of 500 mg/kg body weight. An increase in lung adenomas after 9 months was seen in only one of the 4 strains. It was concluded that the data available were too meagre to allow an evaluation of the carcinogenicity of this organomanganese compound.

8.5 Mutagenicity and Chromosomal Abnormalities

There is little information concerning the mutagenicity of manganese. Processes such as genetic recombination might be affected by manganese through its influence on enzymes that control DNA structure and metabolism. Manganese can be substituted for magnesium in the binding of the two ribosomal subunits as well as in the binding of M-RNA to the whole ribosome (Buttin & Kornberg, 1966). The bone-marrow cells of rats given manganese dichloride at 50 mg/kg body weight, orally, showed an unusual incidence of chromosome aberrations (30.9%) compared with those of control animals (8.5%) (Mandžgaladze, 1966 a,b,c; Mandžgaladze & Vasakidze, 1966). Permanganate given to rats in daily doses of 10 mg/kg body weight, for 9 months, produced an increase in the mitotic activity of bone-marrow cells (Shigan & Vitvitskaja, 1971). Manganous chloride has been reported to be mutagenic for *Escherichia coli* (Demerec et al., 1951; Durham & Wyss, 1957) and *Serretia marcescens* (Kaplan, 1962). Studies on manganese(II,III) oxide (Mn_3O_4) and methylcyclopentadienyl manganese tricarbonyl revealed that neither compound was mutagenic for *Salmonella typhimurium* or *Saccharomyces cerevisiae* (Simmon & Ligon, 1977).

8.6 Miscellaneous Effects

Rats given doses of manganese sulfate of the order of 0.7—2.0 mg showed depressed thyroid activity accompanied by reduced thyroid weight, thinning of follicular epithelium, and smaller follicles (Hakimova et al., 1969). Observations over 6—12 months on rabbits injected with manganese dichloride at 3.5 mg of manganese per kg body weight showed an increase in serum calcium and decreases in serum magnesium and inorganic phosphorus (Jonderko, 1965). Other effects of manganese that have been observed include alterations in immunological activity (Antonova, 1968), disturbance of nitrogen metabolism, (Slavnov & Mandadziev, 1968), and a depression by manganous chloride of the acetylcholine output in the myenteric plexus of guineapigs (Kosterlitz & Waterfield, 1972).

Table 10. Acute toxicity of methylcyclopentadienyl manganese tricarbonyl (MMT) and cyclopentadienyl manganese tricarbonyl (CMT) following a single administration

Compound	Administration route	Animal	LD_{50} (mg/kg)	LC_{50} (mg/m³)	Reference
MMT	oral	mouse	352		Pfitzer et al. (1972)
		rat (male)	175		
		rat (male)	58		Hysell et al. (1974)
		rat (female)	89		Pfitzer et al. (1972)
		guineapig	905		Pfitzer et al. (1972)
		rabbit	95		Pfitzer et al. (1972)
	intravenous	rabbit	6.6		Pfitzer et al. (1972)
	percutaneous (6 h, 10 % peanut oil)	rat	665		Pfitzer et al. (1972)
	(24 h, 10 % kerosene)	rabbit	1350		Pfitzer et al. (1972)
CMT	oral	mouse	150		Arhipova (1963)
		rat	80		Arhipova (1963)
	inhalation (2 h)	rat		120 (LD_{80})	Arhipova et al. (1965)

8.7 Toxicity of Organomanganese Fuel Additives

There are two organomanganese carbonyl compounds that have been considered as gasoline (petrol) additives. In the USSR, cyclopentadienyl manganese tricarbonyl (CMT) has been studied for use as an additive, while, in the USA, the methylated homologue methylcyclopentadienyl manganese tricarbonyl (MMT) has been introduced. In assessing the potential toxicity of the two compounds, both occupational exposure to the parent compounds and exposure of the general population to the combustion products should be considered. Studies on acute toxicity in a number of animal species, following single administrations of MMT and CMT have been summarized in Table 10.

Following oral administration of MMT, rats developed huddling, roughened hair coats, tremors, progressive weakness, laboured respiration, serosanguineous nasal discharge, and terminal coma. All deaths occurred within 6 days. Survivors appeared normal 14 days after exposure. Necropsy findings consisted of saccular atonic stomachs, severe congestion of the liver and lungs, and tubular degeneration in the kidney. The picture differed from acute manganese toxicity, the dose administered being smaller than is needed for acute manganese poisoning in rats, and the liver lesions progressing from acute centrolobular congestion to parenchymal necrosis and, later, extensive cytoplasmic vacuolar changes. Whether the MMT itself or one of its metabolic products was responsible for the toxic effects, could not be assessed (Hysell et al., 1974).

Rats appeared to be more sensitive to MMT and CMT than mice, guineapigs and rabbits, and males seemed more sensitive than females (Arhipova et al., 1963; Pfizer et al., 1972).

Eight out of 20 mice died after 20 days of oral administration of CMT in oil at 25 mg/kg body weight, 6 times weekly. Daily oral administration of 5 mg/kg body weight to rats, for 2 months, only resulted in decreased osmotic resistance of the erythrocytes (Arhipova et al., 1963).

Prolonged inhalation of MMT at a concentration of 15 mg/m³ for 7 h daily, for 150 days, was lethal to mice and rats, whereas a concentration of 6 mg/m³ did not cause any deaths (Pfitzer et al., 1972). Repeated inhalation of CMT at a concentration of 20—40 mg/m³ caused 50% mortality in rats. Prolonged inhalation (10 months) of concentrations of 0.7—1 mg/m³ by rabbits, guineapigs and rats resulted in muscarine effects on the central nervous system, a decrease in diuresis together with an increase in the urinary albumen content, and decreased resistance to infection (Arhipova, 1963).

Three studies have been conducted on the effects of prolonged exposure to the combustion products of MMT. Moore et al. (1975) exposed rats and hamsters for 8 h daily for 56 days to mean concentrations of manganese in air of 131 and 117 µg/m³ in non-irradiated and irradiated exposure chambers, respectively. The general condition and appearance of the experimental animals was not affected during the experiment and no histopathological lesions attributable to manganese exposure were found. However, an increase was noted in manganese tissue concentrations in the exposed rats compared with control rats. In another experiment, rats and monkeys were exposed for 24 h per day over 9 months to combustion products produced by burning MMT vapours in a propane flame. The exposure levels measured as inorganic manganese were 11·6, 112·5, and 1152 µg/m³. Clinical and histopathological investigations performed during, at the end of, and 6 months after the exposure period failed to reveal any adverse effects (Huntingdon Research Center, 1975). In a study by Coulston & Griffin (1976), monkeys were exposed to concentrations of manganese in air of 100 µg/m³ for periods up to 66 weeks without any signs of toxicity. However, manganese levels in tissues increased, particularly in the lungs and pons. Rats exposed in a similar way for periods up to 3 weeks also showed increased manganese concentrations in lung and brain tissue. Two monkeys exposed to about 5000 µg/m³ for 23 weeks failed to exhibit any neurological or behavioural disorders during the exposure period and the following 10-month observation period.

8.8 Mechanisms and Toxic Effects

At present, the relationship between a large number of *in vivo* and *in vitro* effects of manganese cannot be explained in terms of biochemical mechanisms. However, effects on the central nervous system may, to some extent, be explained by recent pathophysiological findings in the brain. Because of its clinical and histochemical resemblance to parkinsonism, it has been possible to associate alterations of the catecholamine metabolism in the brain with extrapyramidal manifestations of manganese poisoning. In parkinsonism, the most constantly affected area of the brain is the substantia nigra, whereas the striatum and pallidum show little damage (Faurbye, 1970; Barbeau et al., 1976). It has been repeatedly shown that the dopamine concentration in the striatum and pallidum is reduced in patients with parkinsonism (Faurbye, 1970), and a causal relationship between cellular loss of the pars compacta of the substantia nigra and depletion of dopamine in the ipsilateral striatum and pallidum has been experimentally demonstrated in monkeys (Poirier & Sourkes, 1965; Sharman et al., 1967; Goldstein et al., 1969), cats (Poirier et al., 1967a), and rats (Faull & Laverty, 1969). In monkeys, this type of brain damage resulted in abnormal motor function on the contralateral side. There was also an ipsilateral decrease in the synthesis of dopamine from the precursor 3,4-dihydroxyphenylalanine (L-dopa) (Poirier et al., 1967b; 1969), the rate-limiting factor of which may be the decreased activity of tyrosine 3-hydroxylase (EC 1.14.16.2) (Levitt et al., 1965; Goldstein et al., 1966). Similarly, it has been shown that lesions at the midbrain level in monkeys were associated with contralateral choreiform movements and depletion of striatal serotonin (Sourkes & Poirier, 1966; Goldstein et al., 1969). In manganese poisoning, the characteristic brain lesions, unlike the lesions in parkinsonism, are in the striatum and pallidum, with little alteration in the substantia nigra (section 9.3.1); nevertheless, in experimental studies on monkeys (Neff et al., 1969) rabbits (Mustafa & Chandra, 1971) and rats (Bonilla & Diez-Ewald, 1974), it was shown that manganese caused depletion of dopamine in the basal ganglia, especially in the striatum. In the study on monkeys, depletion of serotonin was also noted. Moreover, post-mortem biochemical analysis of the brain of a patient suffering from chronic manganese poisoning showed a reduced concentration of dopamine in the striatum and also in the substantia nigra (Bernheimer et al., 1973). These findings implicate the dopaminergic system in the extrapyramidal manifestations of chronic manganese poisoning and this is further supported by the fact that a remarkable improvement in the central nervous system symptoms can be achieved by the administration of L-dopa, a precursor of dopamine (section 9.3.2).

Oxidative enzymes, which are abundant in the pallidum and striatum (Shimizu & Morikawa, 1957), are probably located within the mitochondria (Maynard & Cotzias, 1955) and, thus, are liable to be affected by the accumulation of manganese at these sites. Intact oxidative enzyme systems are needed to supply the energy for the degradation and synthesis of catecholamines involved in synaptic transmission. Any changes in these systems may affect behaviour and could be related to the initial psychiatric phase of chronic manganese poisoning (Mandell & Spooner, 1968). Similarly, tyrosine 3-hydroxylase and other enzymes in the biosynthetic pathway of catecholamines require oxygen (von Euler, 1965), and energy from ATP is needed to transport and compartmentalize essential compounds and to maintain the appropriate membrane and action potentials necessary for neuronal transmission. Any degeneration of neuronal cells would profoundly alter the neural mechanism with consequent clinical effects.

Manganese may also be involved in the interrelationship between biogenic amines and adenosine 3',5'-monophosphate (cyclic AMP). Inhibition by manganese of adenyl cyclase (EC 4.6.1.1) in the membrane of the receptor cell may lead to decreased formation of cyclic AMP and thus uncouple processes that link the interaction of neurohypophyseal hormones with the formation of cyclic AMP. As a result, hormonal action may be inhibited (Bentley, 1967; Cotzias, 1969; Sutherland et al., 1968).

Because of the involvement of metal ions in the neural transmission processes related to neurohormones, manganese concentrations have been determined in various regions of the rat brain. The hypothalamus contained the highest concentration and this may be related to the neuroendocrine function and oestrus disorders observed as a result of manganese deficiency in some species. It may also be required for the proper functioning of glycosyltransferases in the central nervous system (Donaldson et al., 1973).

9. HUMAN EPIDEMIOLOGICAL AND CLINICAL STUDIES

9.1 Occupational Exposure and Health Effects

Manganese exposure is a health hazard in the mining and processing of manganese ores and in the use of manganese alloys in the steel and chemical industries. The majority of cases of manganese poisoning that have been reported have been associated with a combination of high-speed drilling, which produces large amounts of manganese dust, and poor ventilation. However, manganese poisoning can also occur in other types of industry, such as in the production of dry-cell batteries (Emara et al., 1971). Chronic manganese poisoning can result from exposure to high concentrations of manganese dusts. Onset of the disease may occur after only a few months or several years according to the severity of exposure (Ansola et al., 1944b; Rodier, 1955). Damage is reversible, if the patient is removed from exposure at an early stage. However, apparently a sensitivity can develop, since a person who has recovered seems to be prone to contract the illness again.

The signs and symptoms of chronic manganese poisoning have been described in detail by several authors (Flinn et al., 1940; Ansola et al., 1944b; Rodier, 1955; Peñalver, 1955; Schuler et al., 1957; Mena et al., 1967; Letavet, 1973).

According to the severity of the signs and symptoms, poisoning may be divided into 3 stages: (a) A prodromal stage including the generally insidious beginning of the disease, with rather vague symptoms such as anorexia, asthenia, somnolence, insomnia, hyposexuality, and headache; (b) An early clinical stage, when the onset of extrapyramidal manifestations occurs. Speech disturbances are early symptoms in this phase, sometimes leading to muteness. An increased tone of facial muscles results in a mask-like facies and there is also decreased ability to perform skilled movements. Hyperemotionalism is frequent and tendon reflexes in the lower limbs may be exaggerated; (c) Established chronic manganese poisoning, which is characterized by psychomotor disturbances and neurological signs and symptoms. Marked rigidity due to muscular hypertonia, the most conspicuous sign, is most pronounced in the lower limbs and the face. Asthenia, adynamia, muscle pain, paraesthesia, speech disturbances, and disturbances of the libido are typical. The extrapyramidal dysfunction appearing clearly at this stage results in a close resemblance to parkinsonism. However, the tremor is frequently an intention tremor and not resting tremor that is typical of parkinsonism (Klawans et al., 1970). It has been pointed out that, contrary to parkinsonism, manganese intoxication is frequently associated with some degree of dystonia, defined as a postural instability of complementary muscle groups (Barbeau et al., 1976). Psychological signs and symptoms include apathy, unmotivated laughter, a tendency to weep, irritability, restlessness, and hallucinations. Increased salivation and sweating indicate that an autonomic disturbance is also present.

Although manganese levels are elevated in most tissues in chronic manganese poisoning, studies of individuals with well-developed neurological signs and symptoms have revealed serum and blood manganese levels within the normal range. In contrast, healthy miners may have elevated blood manganese levels, suggesting that these indices are of limited value in the diagnosis of chronic intoxication (Mena & Cotzias, 1970).

Another significant aspect of chronic manganese poisoning is marked individual susceptibility (section 9.4), since many miners are exposed to manganese dust but only a small percentage develop manifestations of poisoning (Rodier, 1955; Smyth et al., 1973).

The possible toxic effects of manganese on the lungs were overshadowed during the earlier years by the effects on the central nervous system. It was only after a high death rate from pneumonia in a pyrolusite mill was reported that this association was suspected (Baader, 1932). The cases of pneumonia described by Lloyd Davies (1946) occurred in workers employed in the manufacture of potassium permanganate and exposed to manganese dioxide and, to a lesser extent, to the higher oxides of manganese. The incidence of cases diagnosed as pneumonia over the period 1938—1945 among the men employed (from 40 to 124 men over the 8-year period) averaged 26 per 1000 workers (range 15—63 per 1000) compared with an average of 0.73 per 1000 in a control group. Analysis of dust was performed on 2 days in 1944. Manganese concentrations in air, calculated from the manganese dioxide content of dust, displayed a range, for 6 measurements of 0.1—13.7 mg/m^3. In general, particles were small, 80% being below 0.2 μm. No case of chronic manganese poisoning was detected over the 8-year period. The clinical features of the pneumonia did not differ from conventional pneumonia with the exception that the response to sulfonamide was slow and that the entire respiratory tract from the nose, through the nasopharynx to the alveoli was involved. Permanent pulmonary changes including fibrosis were not observed. As in other workers exposed to chemical irritants in air, pharyngitis was a frequent finding. Mice exposed to dust from the milling room did not show any increased susceptibility to pneumococci types II and IV or to streptococci. The high incidence of pneumonia continued during the period 1946—1948 (Lloyd Davies & Harding, 1949). The primary change was suggested to be an oedematous reaction of the respiratory epithelium resembling chemical pneumonitis, whereas manganese pneumonia was referred to by Rodier (1955) as a condition of acute alveolar inflammation with marked dyspnoea, shallow respiration, and cyanosis.

Suzuki (1970) reported that the incidence of pneumonia in workers in a ferromanganase plant was twice as high as that in another plant situated in the same area.

The laboratory diagnosis of chronic manganese poisoning is non-specific and, at present, there is no adequate diagnostic test, although urinary manganese concentrations may have some value. This measurement, however, does not correlate well with the severity of the clinical signs. Blood levels of manganese provide little clinical information and blood urea nitrogen, fasting blood sugar, enzymes, and electrolytes are usually normal. Rodier (1955) mentions a reduction in excretion of 17-ketosteroids in 81% of his patients, a relative increase in lymphocytes and a decrease in polymorphonuclear leukocytes in 52%, and an increased basal metabolic rate in 57% of patients. Increased haemoglobin values and erythrocyte counts and decreased monocyte counts were reported by Kesić & Häusler (1954) and an increased serum calcium level was observed by Chandra et al., (1974) in 12 cases of manganese poisoning. In mild cases of intoxication, serum adenosine deaminase levels were also elevated. Cotzias (1966) reported that cerebrospinal fluid findings were nonspecific but tended to show slightly increased cell and protein contents.

Most of the cases of manganese poisoning described have occurred in manganese mines. Rodier (1955) reported on 150 cases from Moroccan mines with a total of about 4000 employees. Underground workers engaged in drilling blast holes ran a high risk of developing manganese poisoning; 132 out of 150 cases occurred among such workers. The concentration of manganese in air in the immediate vicinity of rock drilling was reported in one mine to be about 450 mg/m³ and in another, 250 mg/m³.

In a study on 72 Chilean miners exposed to manganese concentrations in air of 62.5—250 mg/m³, 12(16.5%) were found to have neurological disorders. The average exposure time was 178 days, with a range of 49—480 days (Ansola et al., 1944a). A further study on 370 miners exposed to manganese concentrations in air of 0.5—46 mg/m³ showed that 15 workers (4%) had contracted typical manganese intoxication. In these workers, the average time of exposure was 8 years, 2 months, with a range of 9 months—16 years (Schuler et al., 1957).

Flinn et al. (1940) detected 11 cases of manganese poisoning among 34 workers in 2 manganese ore-crushing mills. The highest recorded concentration of manganese in air was 173 mg/m³. The prevalence of manganese poisoning was correlated with both manganese concentrations in air and the duration of employment. No cases were found among 9 workers exposed to concentrations of less than 30 mg/m³. Five out of 6 men exposed for more than 3 years to concentrations exceeding 90 mg/m³ had chronic manganese poisoning.

An occupational health investigation was carried out in Japan in 3 types of industry: a crushing and refining factory, a dry-cell manufacturing plant, and a welding-rod manufacturing plant (Horiguchi et al., 1966; Horiuchi et al., 1970). The results of medical examinations of 134 workers from the 3 establishments were summarized as follows: On neurological examination, signs of disturbances of the central nervous system were clearly observed in 4 refinery workers, and 11 out of 47 refinery workers were suspected of having some neurological disturbances. Four out of 32 persons in the electric welding-rod plant, and 7 out of 55 persons from the dry-cell factory were also suspected of having some form of neurological disturbance (Horiguchi et al., 1966). Horiuchi et al. (1970) reported that a statistically significant correlation existed between the neurological findings and the levels of manganese in the urine of these workers.

The concentrations of manganese in the blood and urine of these workers are shown in Table 11 together with the concentrations in the air of the work area.

Table 11. Manganese concentrations in the blood and urine of mangenese workers in relation to air concentrations in the work area[a]

Type of work	Mn in air (mg/m³)	Mn in whole blood (mcg/100 g)	Mn in urine (mcg/1)	Statistical significance	
				Mn in whole blood	Mn in urine
crushing manganese ore	2.3—17.1 median = 8.4	4—54 median = 9.5	8—165 median = 68.5		
manufacturing dry-cell batteries	1.9—21.1 median = 4.3	4—20 median = 8	1—42 median = 6	$P = 0.0113$[b]	$P = 0.00049$[b]
manufacturing electrodes	3.8—8.1 median = 4.9	4—17 median = 6	3—19 median = 5		

[a] From: Horiuchi et al. (1970).
[b] Statistically significant.

No significant relationship between the length of employment and the concentrations of manganese in blood and urine was found. The manganese concentrations in the blood and urine were higher in the manganese-refining workers than in workers in the other 2 industries, and neurological observations differed significantly. The coefficient of correlation between the manganese quantities in the blood and urine was +0.283 (Horiuchi et al., 1970).

Suzuki et al. (1973a, 1973b) carried out an investigation in 2 ferromanganese factories (factories A and B). The manganese concentrations in air were 4.86 mg/m³ in the mixing and sintering plant and less than 1—2 mg/m³ in other places in factory A. Medical examination of 160 persons revealed that 27 were not healthy, as assessed by a screening questionnaire. More than a third of the 160 workers examined complained of such symptoms as failing memory, fatigue, increased perspiration, and hyposexuality. Thirty-four out of 144 male workers, (24%) exhibited tremor in the fingers, 5 (3,5%), displayed muscle rigidity, 7 (5%), adiadochokinesis and 19 (13%), disturbed balance. The geometric mean manganese concentrations in the blood and urine of 144 out of a total of 160 subjects were 18.4 μg/100 ml and 46 μg/litre, respectively (Suzuki et al., 1973a; 1973c).

In the group of workers that had a blood level of manganese of more than 32.7 μg/100 ml (=geometric mean+1 standard deviation), reduced specific gravity of the whole blood, reduced haemoglobin and haematocrit values, and increased blood pressure, serum GOT, and urinary urobilinogen levels were observed. In the group in which the levels of manganese in the urine exceeded 75 μg/litre (=geometric mean+1 standard deviation), the specific gravity of the whole blood and haemoglobin values were comparatively high (Suzuki et al., 1973c).

The ratio of the concentrations of manganese in urine and blood was positively correlated with both the specific gravity of the whole blood and the haemoglobin value. A negative correlation was found between the urine/blood ratio and the length of a worker's service, the diastolic blood pressure, and serum GOT. In the groups with a low urine/blood ratio, there were many cases of symptoms such as dynamia, failing memory, and hyposexuality (Suzuki et al., 1973c).

In factory B, the concentration of manganese in the air near the electric furnace was 0.6 mg/m^3 before tap, rising to 3.2—8.6 mg/m^3, at tap. An occasional value of 24.3 mg/m^3 was recorded under the conveyor belt, when the crusher was operating. During the medical examinations of 100 electric furnace workers, more than 40% complained of increased perspiration, failing memory, lumbago, footsores, headache, and sleepiness. Furthermore, 8% of the subjects exhibited adiadochokinesis, 10%, finger tremor, and 8%, acceleration of the patellar reflex. However, muscular force, whole blood gravity, and haematocrit values were within the normal range. The geometric means of the concentrations of manganese were 11 μg/100 ml in blood and 45 μg/litre in urine (Suzuki et al., 1973b).

Two investigations of 34 and 199 workers in a manganese steel mill exposed to concentrations of manganese in air in the range of 0.4—15 mg/m^3 showed that manganese levels in blood exceeding 20 μg/100 ml were accompanied by rises in blood cholesterol, total serum lipids, lipoproteins, serum bilirubin, calcium, aminolevulinic acid, and asparagine aminotransferase and decreases in magnesium, haemoglobin, serum proteins, and the glutathione contents of red cells. A review of morbidity and sickness records for a period of 6 years showed a higher incidence of absenteeism and atherosclerosis in exposed workers (Jonderko et al., 1971, 1973a, 1973b, 1974).

A recent epidemiological study of 369 male workers employed in the production of manganese alloys was reported by Šarić et al. (1974). The mean concentration of manganese in air ranged from 0.39 to 16.35 mg/m^3 for the exposed population, while 2 control groups were exposed to concentrations of 4—40 μg/m^3 and 0.05—0.07 μg/m^3 respectively. The data from this study suggest that manganese may contribute to the development of a chronic lung disease. Individuals with a history of smoking appeared to be more affected than nonsmokers and there was a relationship between the degree of smoking and the prevalence of respiratory tract symptoms in the manganese-exposed workers suggesting that smoking may act synergistically with manganese (Šarić & Lučič-Palaić, 1977). A retrospective analysis of absenteeism due to pneumonia and bronchitis, in the same group of workers, revealed that those who were occupationally exposed to manganese were affected by these diseases more frequently than the controls. Data for this study were obtained from medical files covering a 13-year period. Bronchitis was classified into 2 categories: (a) acute and not specifically defined and (b) chronic (Šarić, 1972; 1978). During the epidemiological study, it was noted that manganese-exposed workers, particularly those involved in the production of alloys, had a lower mean systolic blood pressure than the controls. Diastolic pressure was not affected and the lowest mean diastolic pressure was observed in the controls. Factors such as age, body weight, and smoking habits, which may have influenced the results, were taken into acount in their interpretation (Šarić & Hrustić, 1975). A neurological examination performed in the same group of ferromanganese workers showed that 62 out of 369 workers (16.8%) had some neurological signs (Šarić et al., 1977). In most cases, the sign was only a tremor of the fingers (47 workers); 11 workers had pathological reflexes with or without tremor and in 4 workers, cogwheel phenomenon was present as an isolated finding or combined with tremor or pathological reflexes.

Whitlock et al. (1966) reported 2 cases of chronic manganese poisoning, which occurred in a ferromanganese plant where the concentrations of manganese in air were in the range of 0.1—4.7 mg/m³. Examination 4 years later (Tanaka & Lieben, 1969), showed little improvement in one of the cases, the neurological manifestations being unchanged. The other case had improved but walking backward was still a little difficult. His face lacked expression to some extent but was no longer mask-like and the Babinski reflex was unilaterally positive.

In studies by Sabnis et al. (1966), the daily weighted average exposure to manganese was estimated in a ferromanganese plant with 1000 workers. No worker had a weighted average exposure exceeding 2.3 mg/m³. During one year of weekly measurement, the highest recorded manganese concentration in air was 10 mg/m³ and the mean concentrations for various operations ranged from 0.5 to 5 mg/m³. No cases of manganese poisoning were detected in this plant, and when screening for symptoms and signs associated with early manifestation of manganese intoxication, all findings were "almost negative with respect to most of the symptoms".

Weighted average concentrations were also estimated by Smyth et al. (1973), who discovered 5 cases of manganese poisoning among 71 workers studied in a ferromanganese plant. No members of a control group of 71 unexposed workers displayed similar signs and symptoms. Three out of the 5 cases were exposed to manganese fumes and 2 to manganese dust; exposure times varied from 8 to 23 years. One case with 10½ years of exposure to fumes, mainly of Mn(II,III) oxide, was exposed at the time of the investigations to a weighted average concentration of only 0.33 mg/m³, as calculated from 13 air measurements, each sampling period extending over 30 minutes; the highest recorded level was 5.9 mg/m³ and the other 12 measurements were below 5 mg/m³. The patient exhibited facial masking, reduced blinking reflex, micrographia, loss of associated arm movements on the right, tremor of the right hand and some cogwheel rigidity of the right extremities. A high degree of individual susceptibility or additional exposure to manganese seemed to be the likely explanation for this case of poisoning. However, the time of onset of the symptoms in this patient was not discussed, and the beginning of the disease may have been associated with the higher concentrations that prevailed earlier at this plant.

9.2 General Population Exposure and Health Effects

Only one epidemiological report is available on adverse effects from drinking-water contaminated with manganese. Kawamura et al. (1941) studied 16 cases of manganese poisoning, 3 of which were fatal (including one suicide), in a small Japanese community. About 400 dry-cell batteries were found buried within 2 m of a well used as a water supply. The manganese content of the water was about 14 mg/litre and concentrations of 8 and 11 mg/litre were found in 2 other wells. All 16 intoxicated subjects drank water from these wells. The subjects exhibited psychological and neurological disorders associated with manganese poisoning, and high manganese and zinc levels were found in organs at autopsy.

With the introduction of a ferro- and silicomanganese plant in Sauda (Norway), an increase was reported in the incidence of lobar pneumonia in the population living in the vicinity of the plant (Elstad, 1939a; 1939b). During the period 1924—1937, mortality due to lobar pneumonia was 8 times that in the whole country and morbidity was four times higher. Mortality due to lobar pneumonia in the age group 15—39 years was about 20 times that in the whole country and the course of the disease was more severe in Sauda (lethality 35.6%) than in the rest of Norway (lethality 20.3%). The following factors implicated manganese in the etiology of the disease: (a) men working at the factory had a 50% higher mortality due to lobar pneumonia than men employed elsewhere (Elstad, 1939b); (b) there was a positive correlation between morbidity and mortality and the amount of metal produced; and (c) the occurrence and types of pneumococci in Sauda did not differ from the rest of the country (Elstad, 1939a; Riddervold & Halvorsen, 1943). Air analyses were performed in 1930 at a sampling site 3 km downwind from the factory, using a colorimetric assay which involved oxidation of manganese to permanganate. Air was found to contain Mn(II,III) oxide at 30—64 $\mu g/m^3$ and silica (SiO_2) at 6.4—8.9 mg/m³. However, the author stated that the oxidation of manganese to permanganate may have been incomplete, thus yielding rather low results (Bockman, 1939). Elstad (1939a) reported that air samples taken at various places contained levels of manganese oxides ranging from 45 to 64 $\mu g/m^3$. Thus, both reports indicate that manganese levels expressed as the metal may have been about 45 $\mu g/m^3$, at least. Exposure to manganese was further confirmed by the finding of a manganese concentration of more than 150 mg/kg dry weight in the lungs of a woman, who was not working in the factory (Bockman, 1939). Povoleri (1947) also noted that the prevalence of respiratory diseases among the inhabitants of Aosta (Italy) increased with the production of ferromanganese by a plant in that town, but no detailed study of the situation was conducted.

Investigations in Japan include those of Nogawa et al. (1973) and Kagamimori et al. (1973), who studied the health of people living in the vicinity of a ferromanganese plant. The amount of manganese in dustfall that was collected in 4 places in Kanazawa city, far from the factory, averaged 8 kg/km² per month. However, in 3 places, 200—300 m from the factory, the average manganese level was 200 kg/km² per month. There was no difference in the quantities of dustfall and sulfur dioxide in the 2 areas. The 5-day mean concentration of manganese in suspended particulates was 4.04 $\mu g/m^3$ at a point 100 m away from the factory, and 6.70 $\mu g/m^3$ at a distance of 300 m. However, following the smoke downwind, a range of 1-h manganese concentrations of 4.5—260 $\mu g/m^3$ was measured at a distance of 50—700 m from the chimney (Itakura & Tajima, 1972). A comparative study of junior high school students (1258) housed in a school 100 m away from the plant and a similar group (648) housed 7 km away produced the following findings: students in the school 100 m from the factory had a higher prevalence of nose and throat symptoms, a higher prevalence of past-history pneumonia, and a lower pulmonary function (as assessed by measuring the forced expiratory volume, one second capacity, one second ratio, and maximum expiratory flow) than students in the control school. Among the exposed schoolchildren, pulmonary function was lowest in those who had lived in the area longest and in those who lived closest to the factory. A follow-up study, conducted 1 year after a dust collector had been established, showed that the manganese concentrations in suspended particulate matter had decreased by about half at 200—300 m distance from the factory. Furthermore, no differences were found in the symptoms or pulmonary function between exposed and control groups except in third-grade students, who had, presumably, been subjected to long-term exposure and whose lung function showed some deficit. Suzuki (1970) also made observations on pneumonia morbidity in the same area. He found a history of higher rates of pneumonia in schoolchildren and their families near the factory than in a control group.

A 4-year study of the incidence of acute bronchitis, peribronchitis, and pneumonia was carried out on 31 000 inhabitants of a town on the coast of Dalmatia, Yugoslavia, where the atmosphere was polluted by the emissions from a manganese alloy plant. The concentrations of both manganese and sulfur dioxide in the atmosphere were monitored. According to the annual mean concentrations of manganese in air, the town was divided into the following 3 zones: Zone I, 0.27—0.44 $\mu g/m^3$; Zone II, 0.17—0.25 $\mu g/m^3$; Zone III,

0.05—0.07 μg/m³. Since a low-volume sampling technique was used, the particles collected were mainly of respirable size. The concentrations of sulfur dioxide were permanently low, with annual mean levels below 30 μg/m³. The incidence of respiratory diseases was analysed according to zones of manganese exposure and age, sex, and seasonal factors (manganese concentrations usually being higher in the summer than in winter) were taken into consideration. In residential zones, the incidence of acute bronchitis for both sexes was lowest in the zone with the lowest manganese concentration, but the highest incidence did not occur in the zone nearest the factory; however, the concentrations of manganese did not significantly differ in zones I and II. The incidence of pneumonia did not seem to exceed expected values and did not differ significantly in relation to sex, zone, or season. Thus, the expected higher rate of pneumonia in the winter failed to occur, and the authors raised the question as to whether this might be associated with higher summer concentrations of manganese (Šarić, 1978; Šarić et al., 1975). In the evaluation of the study, allowance should be made for the fact that 2—6 times higher concentrations were obtained using a high-volume technique than with a low-volume sampling technique (Šarić, 1978) and that, apart from measuring sulfur dioxide levels, environmental and socioeconomic factors associated with the occurrence of respiratory illness were not taken into consideration.

A study in the USSR carried out on 928 wives of workers employed in various manganese-processing plants showed that 13.8% had spontaneous abortions and 3.2% had stillbirths, while in a matched control group the incidence of these disorders was 8.1% and 1.7%, respectively. The rate of spontaneous abortion appeared to increase with the duration of exposure of the workers. Thus, the wives of workers employed for 10—20 years had a spontaneous abortion rate of 15.4% while the wives of workers exposed for 5—10 years had a spontaneous abortion rate of 11.7% (Mandžgaladze, 1967). No information regarding the type of work of the wives was given.

9.3 Clinical Studies

9.3.1 Pathomorphological studies

There have been relatively few documented autopsy reports concerning pathological changes in man related to manganese poisoning (Casamajor, 1913; Ashizawa, 1927; Canavan & Drinker, 1934; Stadler, 1936; Voss, 1939, 1941; Flinn et al., 1940; Kawamura et al., 1941; Parnitzke & Peiffer, 1954). In the central nervous system, the most extensive changes have been found in the striatum (caudate nucleus and putamen) and pallidum. Ashizawa (1927) noted a loss of ganglion cells in the pallidum and marked degeneration of the putamen and caudate nucleus; he also reported slight changes in the substantia nigra. Brain atrophy over the vertex and lateral aspects was reported by Canavan & Drinker (1934) in a patient who died 14 years after the onset of symptoms. On the frontal section, the atrophy was conspicuous and dilatation of the lateral ventricles with shrinking of the basal ganglia was found. Degeneration of nerve cells was seen in the basal ganglia together with gliosis and satellitosis. The caudate nucleus, putamen, globus pallidus, and thalamus were equally affected. Only diffuse cellular changes were seen in the cerebral cortex and cerebellum. Kawamura et al. (1941) reported that a 46-year-old patient died one month after the onset of an illness contracted from drinking-water heavily contaminated with manganese. Moderate congestion was noted in the brain, spinal cord, and meninges, with meningeal oedema most prominent in the occipital part. Severe degeneration was found in the globus pallidus, whereas the thalamus, caudate nucleus, and Louis' body were histologically normal and no increase in glial cells was found. The most pronounced changes found by Stadler (1936) were in the striatum and pallidum which were equally affected. Perivascular degeneration with loss of ganglion cells and proliferation of glial cells were typical findings in the putamen and the caudate nucleus. Less severe changes were found in the cortex and only slight changes in thalamus, hypothalamus, and cerebellum.

Generalized atrophy of the liver cell cords, most marked at the centre of the lobules, was found by Flinn et al. (1940). However, because of post-mortem changes, the findings on examination of the brain remained obscure. The 2 cases described by Voss (1939, 1941) both displayed degenerative changes of the pyramidal tract, whereas histopathological autopsy findings in the striatum, pallidum, caudate nucleus, putamen, and the cortex were minimal. Degeneration of peripheral nerves was present in both cases: however, one of the patients suffered also from amyotrophic lateral sclerosis (Voss, 1939). According to Parnitzke & Peiffer (1954), a 19-year-old manganese miller, who developed symptoms and signs after 1 year of exposure, died 24 years later after progressive impairment of neurological function. The total exposure time was approximately $2\frac{1}{2}$ years. A loss of ganglia cells in the pallidum with glial cell proliferation and high concentrations of manganese, lead, and iron in the plexus chorioideus were the major findings at autopsy.

Bernheimer et al. (1973) described morphological findings in a woman with chronic manganese encephalopathy, who died with the clinical picture of a rigid-akinetic parkinsonian syndrome. She had been working for many years in a battery factory, and, 10 years after giving up this work, she had blood manganese levels 10 times higher than reference values. Generalized astroglial proliferation was found with a preference for certain cortical areas, the putamen, pallidum, and red nucleus. There was also mild pallidal atrophy and marked degeneration in the zona compacta of the substantia nigra. An interesting finding was the low concentrations of dopamine in the striatum and of noradrenaline in the hypothalamus; however, serotonin levels were considered to be normal.

9.3.2 Therapeutic studies

Treatment of chronic manganese poisoning has recently undergone a basic change reflecting a better understanding of the pathophysiology of the condition. Early attempts using various chelating agents, particularly EDTA, were of conflicting benefit but did seem to produce some improvement in the condition, if applied in its early phase, when presumably neuronal destruction had not yet occurred. No improvement could be expected after structural neurological injury had occurred. The results of Penalver (1955) and Tepper (1961) confirmed this and they regarded the treatment as ineffective. Whitlock et al. (1966) reported that treatment with intravenous calcium EDTA mobilized body deposits of manganese (as evidenced by increased levels of manganese in the urine) and led to improvement in muscle strength and coordination within 2—3 months of treatment; however, a follow-up of the 2 cases, 4 years later, revealed that the improvement was persistent in only one case and that the other had deteriorated (Tanaka & Lieben, 1969). That the improvement following EDTA treatment might be temporary was also reported by Cook et al. (1974). Successful treatment has been reported, particularly for early manifestations of manganese poisoning, using calcium EDTA and other chelating agents derived from polyamino-polyphosphoric acid (Mihajlov et al., 1967; Arhipova et al., 1968). Wynter (1962) reported poor results with EDTA in 7 cases in the advanced phase but encouraging results in one patient with early signs and symptoms. Similarly, Smyth et al. (1973) found EDTA treatment successful in 2 cases, both of which displayed loss of associated arm movement as the only neurological sign, but no improvement in 3 cases with more advanced neurological signs.

The essentially negative results with chelating agents may be explained by the fact that increased levels of tissue manganese, that would be amenable to such treatment, were found only in healthy, actively working miners. Apparently, crippled ex-miners had cleared the manganese loads they once had, but did not show any improvement in their neurological picture, indicating that neurological signs can persist in the absence of elevated tissue concentrations (Cotzias et al., 1968). Thus, chelating agents can hardly be

expected to have beneficial effects except in early cases, as no tissue
concentrations remain to be cleared in the later stages of the disease.

Recognizing that a similar biochemical defect was present in
parkinsonism (section 8.6), Mena et al. (1970) used oral doses increasing up to 8.0 gm per day of the dopamine precursor 3-hydroxy-
L-tyrosine (L-dopa) in 6 patients. Five subjects showed reduction or
disappearance of rigidity and hypokinesia, and regained their sense
of balance. The sixth patient displayed aggravation of neurological
signs during L-dopa treatment but responded favourably to a daily
dose of 3 gm of 5-hydroxytryptophane, a precursor of serotonin.
The rational basis for the use of 5-hydroxytryptophane was that
muscle hypotonia, sometimes present in chronic manganese poisoning, but hardly ever present in parkinsonism (Cotzias, 1969), was
probably related to low striatal serotonin levels. This is the case
of the hypotonia of Down's syndrome, which can be reversed by the
administration of the serotonin precursor (Bazelon et al., 1967).
Experimental support for the use of dopamine or serotonin precursors in manganese poisoning includes the fact that the administration of manganese to rats, rabbits, and monkeys was followed by
depletion of striatal dopamine and serotonin (Neff et al., 1966;
Mustafa & Chandra, 1971; Bonilla & Diez-Ewald, 1974) and that
administration of dopamine and serotonin precursors resulted in an
increase in the striatal concentrations of both dopamine and serotonin (Poirier et al., 1967b; Goldstein et al., 1969; Neff et al., 1969;
Bonilla & Diez-Ewald, 1974). However, data on the depletion of serotonin are still conflicting (Goldstein et al., 1969). A favourable result
using L-dopa in one case of chronic manganese poisoning was also
obtained by Rosenstock et al. (1971), but the beneficial response to
L-dopa could not be confirmed by Cook et al. (1974), who treated
3 patients with the drug. It has been proposed that the beneficial
therapeutic effect of L-dopa depends on the dopaminergic fibres not
being completely degenerated (Goldstein et al., 1969; Mena et al.,
1970). The therapeutic doses used in chronic manganese poisoning
have generally been well tolerated, although doses of up to 12 g of
L-dopa have been administered daily (Rosenstock et al., 1971). Considering the high doses needed, Cotzias (1969) drew attention to the
possibility of choline or methionine deficiency resulting from the
donation of their methyl groups, which are needed for the metabolism of L-dopa.

9.4 Susceptibility to Manganese Poisoning

Several authors have tried to explain the individual susceptibility
of miners to chronic manganese poisoning on the basis of nutritional
deficiencies and variations in intestinal absorption. Altstatt et al.

(1968) showed that manganese and iron metabolism are closely related, and Mena et al. (1969) reported that individuals with increased intestinal iron absorption had an increased absorption of manganese as well. Thus, an intestinal absorption of manganese of 7.5% was found in anaemic subjects compared with 3% in healthy subjects. The accelerated turn-over of manganese with a concomitant increase in iron excretion found in heavily exposed workers with elevated tissue levels of manganese (Section 6.1.2.1) may further aggravate a pre-existing anaemia (Mena et al., 1969). Moreover, Mena et al. (1974) reported that the binding capacity of the plasma of anaemic rats was more than twice that of healthy rats. The authors considered that the increase in the transport capacity of the plasma to the blood-brain barrier was related to the finding of about 100% higher concentrations in the brain of anaemic rats.

It was also found that the penetration of the blood-brain barrier in newborn and infant rats less than 18 days old was 4 times that of adult rats (Mena et al., 1974). According to Mena (1974), young rats had an intestinal absorption of 70% of manganese compared with 1—2% in the adult rat. Human data on these aspects are not available.

9.5 Interaction

Apart from individual susceptibility, some individuals may be at higher risk because of exposure to certain chemical or physical factors that may influence the toxicity of manganese. A recent study on occupationally exposed workers (section 9.1) indicated that smoking may act synergistically with manganese in the development of nonspecific respiratory disorders (Šarić & Lučič-Palaić, 1977). Combined exposure to manganese and vibrations or X-rays increased the toxic effects of manganese, particularly in the central nervous system and the adreno-cortical system (Mihajlov et al., 1969; Levanovskaja & Neizvestnova, 1972; Neizvestnova, 1972a, b; Počasev & Neizvestnova, 1972). An increase in the toxicity of manganese compounds has been noted with exposure to chemicals such as carbon monoxide, silicon dioxide, sulfur dioxide, fluorine, copper, and lead (Belobragina & El'nicnyh, 1969; Belobragina et al., 1969; Davydova, 1969; El'nicnyh, 1969; Mihajlov et al., 1969; Davydova et al., 1971; Rylander et al., 1971; Belobragina, 1972; Mavrinskaja et al., 1972; Rylander & Bergström, 1973).

An inhibiting effect of manganese on muscle tumorigenesis caused by nickel subsulfide in rats was reported by Sunderman et al. (1974). Manganese ions were also found to antagonize the excitation of myocardial fibres in frogs caused by nickel ions (Babskiji & Donskih, 1972; Donskih & Mukumov 1974).

10. EVALUATION OF HEALTH RISKS TO MAN FROM EXPOSURE TO MANGANESE AND ITS COMPOUNDS

10.1 Relative Contributions of Air, Food, and Water to Total Intake

10.1.1 General population

In areas without manganese-emitting industries, the annual average concentrations of manganese in air are usually in the range of 0.01—0.07 $\mu g/m^3$. In areas with major foundry facilities, concentrations of about 0.2—0.3 $\mu g/m^3$ can be expected. When ferromanganese or silicomanganese industries are present, the annual average concentration in the surrounding areas may increase to over 0.5 $\mu g/m^3$; occasionally, annual values up to 8.3 $\mu g/m^3$ have been recorded. Assuming a standard respiration rate of 20 m^3 per day, the daily intake of manganese through inhalation in unpolluted areas would be below 2 $\mu g/day$, whereas in the presence of ferromanganese and silicomanganese industries, in extreme situations, the daily mean intake may increase to over 150 $\mu g/day$. Thus, in most instances, the daily intake through inhalation constitutes less than 0.1% of the total daily intake, and rarely exceeds 1%, even in heavily polluted areas.

There is no information on the rate of absorption of inhaled manganese particles. The particle size with which airborne manganese is associated is usually within the respirable range. Inhaled manganese particles are partly cleared through pulmonary defence mechanisms and swallowed. The small size of the particles favours a widespread airborne distribution of manganese, that reaches man indirectly, as a result of fallout on soil and water and through uptake by plants and animals.

The mean levels of manganese in drinking-water are usually about 5—25 $\mu g/litre$, but individual samples from municipal supplies have shown concentrations ranging from trace levels to 100 $\mu g/litre$. Values one order of magnitude higher have been determined in certain rivers. Assuming a standard daily water consumption of 500—2200 ml (ICRP, 1975), the average daily intake of manganese with water is in the range of 2—55 μg and is unlikely to constitute more than 1—2% of the total intake of manganese. Data on the form of manganese present in drinking-water and on the rate of absorption from the gastrointestinal tract of manganese in water are not available.

Most foodstuffs contain manganese in concentrations below 5 mg/kg (wet weight). Grain, rice, and nuts may have manganese levels exceeding 10 mg/kg, whereas finished tealeaves contain several hundred mg/kg. The daily intake of manganese with food, by adults, is about 2—9 mg. In young children and up to adolescence, the intake is about 0.06—0.08 mg/kg body weight and in breastfed and bottlefed infants is as low as 0.002—0.004 mg/kg body weight. The small amount of information available indicates an absorption from the human gastrointestinal tract of less than 5% in healthy adults. The low absorption is supported by studies on mice and rats, in which absorption may range from 0.2 to 3%. There is no information available concerning absorption in infants and young children, but animal data indicate that it could be significantly higher than in adults. A gastrointestinal absorption of 5% would result in an absorbed dose of 100—450 μg/day for adults. The chemical forms and the possible differences in biological availability of manganese present in various foods are not known.

10.1.2 Occupationally-exposed groups

Recent data on the levels of manganese in air in manganese mines were not available to the Task Group, but most studies indicate that levels of several hundred milligrams per cubic metre may occur. Values ranging from 0.8 to 17 mg/m³ have been reported in ore-crushing plants. In steel plants, air concentrations are generally in the range of 0.1—5 mg/m³, only rarely exceeding 10 mg/m³. However, welders may be exposed to air concentrations exceeding 10 mg/m³. A major part of manganese present in the air in work areas is in the form of oxides, and data concerning other compounds are not available. Manganese in dust and fumes appears to be predominantly associated with particles below 5 μm.

10.2 Manganese Requirements and Deficiency

The daily required intake of manganese for adult man appears to be 2—3 mg, and taking into account available data, an estimated minimum daily intake of 1.25 mg would seem adequate for pre-adolescent children. Although newborn infants display a negative manganese balance during the first weeks of life, as manganese is excreted from tissue stores accumulated during fetal life, signs of manganese deficiency have not been seen. Manganese deficiency has been described only once in man in connexion with experimentally induced vitamin-K deficiency and the accidental omission of manganese from the diet. All dietary studies on daily manganese intake have indicated that the daily requirements mentioned earlier are met. Furthermore, it seems that regulatory absorption and excretion mechanisms exist which make manganese deficiency unlikely in man.

10.3 Effects in Relation to Exposure

The primary target organs of inhaled manganese are the lungs
and the central nervous system, although effects have occasionally
been noted in other organs. The effects of manganese are not specific
and a suitable biological indicator of the absorbed dose of manganese
has not been identified, so far. In the measurement of inhalation
exposures, personal samplers have rarely been used. Most data
have been derived from occasional measurements at fixed sampling
sites and do not necessarily represent actual exposure. Thus, exposure-effect and exposure-response relationships cannot be established for manganese on the basis of existing data. The scantiness
of retrospective exposure data makes it difficult to associate effects
with any long-term exposure levels.

10.3.1 General population

With a few exceptions, effects on the central nervous system
have only been observed in occupationally exposed individuals.
However, one incident involving 16 cases of severe manganese
poisoning from drinking-water (manganese concentrations ranging
from 8 to 14 mg/litre) contaminated by discarded dry-cell batteries
indicates the importance of ensuring the proper disposal of manganese-containing wastes.

An increased incidence of pneumonia and nonspecific effects on
the respiratory tract have been reported in populations living in
areas associated with manganese alloy production. In Sauda,
Norway, a 4-fold increase in morbidity and an 8-fold increase in
mortality due to lobar pneumonia were noted over a 14-year period,
in a community living in the vicinity of a ferromanganese and
silicomanganese plant. Moreover, the morbidity and mortality due to
lobar pneumonia varied with the amount of manganese alloy produced by the plant. Results of bacteriological investigations in Sauda
were similar to those in the rest of the country, which further indicates that manganese played at least some part in the etiology
of the pneumonia. The actual exposure of the population to manganese oxide in air is uncertain in this case as only one value (64
$\mu g/m^3$) was given and it was later proved that the analytical method
used gave rather low results. Another study, from Aosta, Italy,
reporting an increase in mortality due to pneumonia as production
of manganese alloys increased at the local ferromanganese plant
is even more difficult to evaluate as no detailed investigations were
carried out.

A higher prevalence of nose and throat symptoms, anamnestic
pneumonia, and decreases in pulmonary function were registered

in a group of schoolchildren living about 100 m from a ferromanganese plant, compared with children attending a school 7 km away
from the factory. The manganese concentrations in air in the polluted
area ranged from 4.0 to 6.7 μg/m^3, (5-day mean values). However, on
3 occasions, short-time samples (1-h) following the smoke downwind
exceeded 100 μg/m^2 at distances of 50—700 m from the chimney, with
a maximum value of 260 μg/m^3. The fact that a follow-up study,
carried out after a dust collector had been in operation for 1 year,
failed to show similar differences between more exposed and less
exposed groups further incriminates manganese in the etiology of
the respiratory findings.

In a study in Yogoslavia, health effects in 2 populations (8700 and
17 100 individuals), exposed to mean annual concentrations of manganese in air of about 0.27—0.44 and 0.16—0.24 μg/m^3, respectively,
arising from a manganese alloy plant, were compared with those in
a population exposed to concentrations below 0.1 μg/m^3. An increased incidence of bronchitis was noted in the exposed populations,
but the incidence of pneumonia did not exceed expected values. The
exposures may have been considerably higher owing to the low-
volume sampling technique used. Although sulfur dioxide concentrations were measured, other environmental and socioeconomic
factors, not considered, might have influenced the results.

10.3.2 Occupationally-exposed groups

Confounding factors have been reported in all the available
studies relating effects to occupational manganese exposure. Some
studies have related certain effects of manganese (including
pneumonia, effects on the central nervous system, and subjective
symptoms) to levels of exposure. As there are very few data concerning retrospective exposure, and subtle neurological and psychological symptoms and signs may have existed unrecognized for
several years, the development of effects may, in fact, have been
related to earlier, higher exposures. Moreover, the relationship
between the development of signs and symptoms and short-term exposure to high concentrations is not known. Although these considerations should be borne in mind in the assessment of exposure-effect
relationships, sufficient information exists to relate at least some
effects to a range of manganese concentrations in air.

10.3.2.1 *Effects on the central nervous system*

Signs and symptoms of extrapyramidal disorders, characteristic
of manganese poisoning, were reported in 2 manganese steel workers
engaged in arc-burning. Measurements of manganese concentrations

in air ranged up to 4.7 mg/m³. At re-examination 4 years later, one worker showed further neurological deterioration, and the other still displayed slight neurological signs.

In a ferromanganese plant, manganese concentrations in air ranged from 1.9 to 4.9 mg/m³ in the sintering area and were below 2 mg/m³ in other areas of the plant. Examination of 160 workers revealed symptoms such as failing memory, fatigue, increased perspiration, and hyposexuality in 30% of the subjects. Of 144 workers, 24% exhibited tremor of the fingers, 13% disturbed balance, 5%, adiadochokinesis, and 3.5%, muscular rigidity. In another ferromanganese plant, exposure levels were mainly below 1 mg/m³, and manganese concentrations up to 3.2—8.6 mg/m³ were measured around the electric furnace at tap only (one value of 24.4 mg/m³ was measured under a conveyor belt that was in operation). Among the 100 workers, 40% complained of symptoms but not all the symptoms were necessarily due to manganese exposure. Adiadochokinesis was found in 8% of the workers, tremor of the fingers in 10%, and eccelerated patellar reflex in 8%. Similarly, slight neurological signs were found in 7 out of 55 dry-cell battery plant workers exposed to a median manganese concentration in air of 4.3 mg/m³ (range 1.9—21.1 mg/m³).

One report from a ferromanganese plant with 1000 employees indicated annual mean concentrations for various operations ranging from 0.5 to 5 mg/m³; the highest concentration recorded was 10 mg/m³, and the highest daily weighted average exposure for any worker was 2.3 mg/m³. The plant physician's register did not show any complaints that suggested manganese poisoning and a screening for subjective symptoms was negative. Unfortunately, clinical examinations were not carried out on the workers and reporting on symptoms was inadequate.

In this context, it is pertinent to consider the fact that characteristic central nervous system signs were produced in monkeys exposed to 0.6—3.0 mg of manganese dioxide per m³ of air, for 1 h per day, over a 4-month period.

10.3.2.2 Manganese pneumonia

Considering that a causal connexion between exposure to manganese and pneumonia has been repeatedly suggested in the literature since 1921, there have been surprisingly few studies concerned with the relationship between the incidence of pneumonia and the type and level of exposure. However, a 35-fold increase in the incidence of pneumonia was reported in workers engaged in the manufacture of potassium permanganate. The incidence was 26 per 1000 workers compared with 0.73 in a control group. The manganese concentrations in air ranged up to 14 mg/m³, as calculated from measurements of manganese dioxide in dust. Although measurements were scarce and higher concentrations may have existed in the beginning of the 8-year period of the study, the fact that no signs of chronic manganese poisoning were observed suggests that the exposure was comparatively low. It is not possible to conclude whether manganese exerts a direct chemical action on the lungs or causes an increased susceptibility to bacterial or viral agents.

10.3.2.3 *Nonspecific effects on the respiratory tract*

A recent epidemiological study of 367 male workers, exposed to a mean concentration of manganese in air of up to 16.4 mg/m³, indicated that manganese may contribute to the development of chronic bronchitis. The higher rate of symptoms related to the respiratory tract in smokers from the exposed group compared with smokers in the control group suggested that smoking may act synergistically with manganese. Retrospective studies of absenteeism because of respiratory disorders have also indicated that populations occupationally exposed to manganese are more frequently affected by these conditions than unexposed populations.

In the assessment of exposure-effect levels in occupational health, it may be useful to consider that exposure of rats to manganese dioxide at 0.3 mg/m³, by inhalation, caused inflammatory changes in the respiratory tract of the animals, and that inhalation of the same compound at 0.7 mg/m³ and 3.0 mg/m³ caused mottling of the pulmonary X-rays of monkeys as well as inflammatory alterations in the respiratory tract of mice.

10.3.2.4 *Diagnosis of manganese poisoning and indices of exposure*

The clinical diagnosis of manganese poisoning may be difficult, particularly in the early stages of the disease, since reliable diagnostic procedures are not available. Urine and blood manganese levels are only weakly correlated with the degree of exposure and with the severity of the toxic response. Pulmonary manifestations may be absent and when present in smokers, for instance, they may easily be ignored. The onset of psychological and neurological signs and symptoms is often insidious, and the manifestations nonspecific, whereas later stages resemble parkinsonism. Apart from the ascertainment of exposure, repeated screening for subjective symptoms and thorough neurological examinations, together with the determination of the manganese concentrations in urine and blood appear, at present, to be the only methods available for the detection of the disease. Since manganese is eliminated primarily in the faeces, the estimation of faeces manganese may serve as a useful guide to exposure, although this approach has rarely been applied.

10.3.2.5 *Susceptibility and interaction*

The incidence of chronic manganese poisoning among workers exposed to high manganese concentrations has shown highly variable individual susceptibility to the effects of manganese. The reasons for higher susceptibility in some individuals are not clear though the close relationship between manganese and iron metabolism may afford one explanation. The intestinal absorption of manganese in anaemic subjects is twice that of healthy individuals. Moreover, exposed workers with elevated tissue levels of manganese have an increased excretion of manganese combined with a concomitant increase in iron excretion, which may aggravate a pre-existing anaemia. Animal experiments have shown that the binding capacity of the plasma of anaemic rats is more than double that of healthy rats and that the entrance of manganese into brain is higher in anaemic rats. Penetration into the brain of newborn and infant rats is 4 times that of adult rats and the intestinal absorbtion of manganese in young rats may be up to 70% compared with 1—2% in adult rats.

There is little information on the interaction between manganese and other chemical and physical factors. However, some studies have indicated that manganese, smoking, and sulfur dioxide may produce synergistic effects on the respiratory tract. Vibration and X-rays have been reported to increase the toxic effects of manganese on the central nervous system and this may also be true of other chemicals such as carbon monoxide, silicon dioxide, fluorine, copper, and lead.

10.4 Organomanganese Compounds

Two classes of organomanganese compounds should be considered from the toxicological point of view. One includes manganese ethylene-bis-dithiocarbamate (Maneb), a fungicide used on edible crops. Here, the manganese entity is of little toxicological significance, while the organic fraction is part of a larger problem concerning the use of this class of fungicides. These fungicides have been considered by a joint FAO/WHO Meeting on Pesticide Residues in food (WHO, 1969). The International Agency for Research on Cancer recently considered data relevant to the carcinogenicity of manganese ethylene-bis-dithiocarbamate and concluded that the data available were too meagre for an evaluation of the carcinogenic risks of this compound man (IARC Working Group, 1976).

The second class of organomanganese compounds of potential toxicological significance is constituted by the manganese tricarbonyl compounds used as additives in petrol. As only a small portion of the parent compound is emitted and this is rapidly photodecomposed to mainly unknown compounds, exposure to manganese tricarbonyl compounds is more likely to constitute an occupational hazard than a public one. Widespread use of these compounds as petrol additives will, however, result in increased exposure of the general population to the combustion products, mainly inorganic manganese, and will especially affect urban environments. Animal experiments have not revealed any adverse effects from the long-term exposure (up to 66 meeks) of rats, hamsters, and monkeys to combustion products of methylcyclopentadienylmanganese tricarbonyl with manganese concentrations in air in the range of 12—5000 $\mu g/m^3$. However, at 100 $\mu g/m^3$ for up to 66 weeks, the tissue levels of manganese increased significantly in monkeys. At present, data on the effects of prolonged exposure of man to low concentrations of manganese in air and on the effects of combined exposure to manganese and other pollutants are inadequate for an evaluation of the health risks, if any, that may arise from an substantial increase in the use of manganese tricarbonyl compounds in petrol. Studies on the effects on exhaust gas emissions of manganese tricarbonyl compounds as additives in petrol are also, in some respects, conflicting and further studies are needed.

10.5 Conclusions and Recommendations

Manganese is an essential trace metal for both man and animals. Manganese deficiency is extremely unlikely to occur in man because there is a sufficient supply of manganese in the diet and because of homeostatic mechanisms present in man.

Systemic effects from over-exposure to manganese, which constitute an inhalation hazard for occupationally-exposed populations, may occur in other populations, but only in cases of accidental or intentional ingestion of exceptional amounts of the metal. Pneumonia and nonspecific effects on the respiratory tract may occur both in occupationally-exposed populations and in the general population, in areas associated with industrial emissions of manganese.

The assessment of health risks related to both occupational and community exposure to manganese is made more difficult by the generally poor quality of available information particularly in the case of exposure data.

10.5.1 Occupational exposure

Signs and symptoms of effects on the central nervous system may already occur at air concentrations of manganese in the range of 2—5 mg/m³. A minimum exposure-effect level cannot be assessed, but considering human and animal data as well as the highly variable individual susceptibility to manganese, it is probably less than 1 mg/m³.

Exposure-effect relationships for pneumonia and nonspecific respiratory effects cannot be established from available occupational data. Animal data indicate a local effect of manganese dioxide on the respiratory tract at concentrations ranging from 0.3 to 3.0 mg/m³. It seems possible that characteristics such as particle-size distribution and type of manganese compound are etiologically more important than mass concentrations of manganese. Special attention should be paid to the possibility of concomitant exposure to other pollutants which may act synergistically with manganese on the respiratory tract. There is a conspicuous discrepancy between, on the one hand, extremely low concentrations of manganese in the ambient air reported to cause effects on the respiratory tract, and, on the other hand, the scantiness of reports of similar effects in populations occupationally-exposed to about 100—1000 times these levels.

Persons with psychological or neurological disorders are not suitable for work associated with exposure to manganese.

Nutritional deficiency states may predispose to anaemia, increasing susceptibility to manganese; and subjects suffering from such deficiencies should be under surveillance.

In the absence of specific diagnostic means, the worker should be screened for subjective symptoms and subjected to clinical examinations at regular and not too long intervals. A pre-employment examination is clearly needed.

10.5.2 General population exposure

At present, there is no evidence that the manganese concentrations of less than 0.1 μg/m³ generally found in ambient rural and urban air are associated with any health risk to man.

Annual mean concentrations of manganese in air exceeding 0.1 μg/m³ are invariably man-made and are found in areas associated with manganese-processing industries. Manganese compounds may be widely used as petrol additives in the future and may cause urban air concentrations of manganese to exceed this level.

Increased morbidity and mortality due to pneumonia, and nonspecific effects on the respiratory tract in the general population have been related to increased exposure caused by nearby manganese alloy plant. The documentation available is inadequate for the establishment of guidelines with respect to manganese concentrations in ambient air.

In view of existing data and considering the possibility of increasing use in the future of organomanganese compounds as petrol additives, it is recommended that epidemiological surveys be conducted in communities exposed to annual mean concentrations of manganese in air exceeding 1 μg/m³.

REFERENCES

ABDULLAH, M. I. & ROYLE, L. G. (1972) Heavy metal content of some rivers and lakes in Wales. *Nature (Lond.)*, **23**: 329—330.

ABRAMS, E., LASSITER, J. W., MILLER, W. J., NEATHERY, M. W., GENTRY, R. P., & SEARTH, R. D. (1976) Absorption as a factor in manganese homeostasis. *J. anim. Sci.* **42**: 630—636.

AJEMIAN, R. S. & WHITMAN, N. E. (1969) Determination of manganese in urine by atomic absorption spectrometry. *Am. Ind. Hyg. Assoc. J.*, **30**: 52—56.

AKSELSSON, R., JOHANSSON, K., MALMQVIST, K., FRISMARK, J., & JOHANSSON, T. B. (1975) Elemental abundance of variation with particle size in aerosols from welding operations. In: *Proceedings of the Conference on Nuclear Methods in Environmental Research, Columbia, Missouri July 29—31*, 1974, 7 pp.

ALEXANDER, F. W., CLAYTON, B. E., & DELVES, H. T. (1974) Mineral and trace-metal balances in children receiving normal and synthetic diets. *Q. J. Med.*, **169**: 89—111.

ALJAB'EV, G. A. & DMITRIENKO, M. M. (1971) [Trace elements in edible and poisonous fungi in the Irkutsk Oblast.] *Nauk. Tr. Irkutsk. med. Inst.*, **107**: 90—94 (in Russian).

ALTSTATT, L. B., POLLACK, S., FELDMAN, M. H., REBA, R. C., & CROSBY, W. H. (1968) Liver manganese in hemachromatosis. *Proc. Soc. Exp. Biol. Med.*, **124**: 353—357.

AMDUR, M. O., NORRIS, L. C., & HEUSER, G. F. (1945) The need for manganese in bone development in the rat. *Proc. Soc. Exp. Biol.*, **59**: 254—255.

AMERICAN CONFERENCE OF GOVERNMENTAL INDUSTRIAL HYGIENISTS (1958) Determination of manganese in air.: Periodate oxidation method. In: *Manual of analytical methods recommended for sampling and analysis of atmospheric contaminants*, Cincinnati, OH, American Conference of Governmental Hygienists, pp. MN-1 to Mn-4.

ANGELIEVA, R. (1969) [Spectrographic method for determination of manganese in foodstuffs of vegetable origin by emissive spectral analysis.] *Hig. Zdraveop.*, **3**: 312—320 (in Bulgarian).

ANGELIEVA, R. (1970) [Method for determination of the quantity of manganese in foodstuffs of vegetable origin by emissive spectral analysis.] *Hig Zdraveop.*, **3**: 312—320 (in Bulgarian).

ANGELIEVA, R. (1971) [Spectrographic method for determination of the manganese content of urine.] *Hig. Zdraveop.*, **5**: 517—520 (in Bulgarian).

ANKE, M. & SCHNEIDER, H.-J. (1974) [Trace element concentrations in human kidney in relation to age and sex.] *Zschr. Urol.*, **67**: 357—362 (in German).

ANSOLA, J., UIBERALL, E., & ESCUDERO, E. (1944a) [Intoxication by manganese in Chile (Study on 64 cases). I. Environmental and etiological factors.] *Rev. Med. Chile*, **72**: 222—228 (in Spanish).

ANSOLA, J., UIBERALL, E., & ESCUDERO, E. (1944b) [Intoxication by manganese in Chile (Study on 64 cases). II. Clinical aspects, incapacity and medico-legal reparation.] *Rev. Med. Chile*, **72**: 311—322 (in Spanish).

ANTONOVA, M. V. (1968) [The action of the trace element manganese contained in the food ration on the immunobiological reactivity of the organism.] *Vopr. Pitan.*, **3**: 36—41 (in Russian).

ANTOV, G. ZLATEVA, M. (1974) [On the effect of lead and manganese upon the connective tissue of the aorta in albino rats.] *Farmakol. Toksikol.*, **37**: 96—98 (in Russian).

ARHIPOVA, O. G. (1963) [On the mechanism underlying the action of

cyclopentadienyltricarbonyl manganese — a new antiknock compound.]
Gig. Tr. prof. Zabol., **7**: 43—49 (in Russian).

ARHIPOVA, O. G., TOLGSKAJA, M.S., & KOČETOVA, T. A. (1963) [Toxic
properties of manganese cyclopentadienyltricarbonyl antiknock sub-
stance.] *Gig. i Sanit.*, **28** (4): 29—32 (in Russian).

ARHIPOVA, O. G., TOLGSKAJA, M. S. & KOČETKOVA, T. A. (1965) [The
toxicity of vapours of a new antidetonator — manganese cyclopentadienyltricarbonyl in the air of industrial premises.] *Gig. i Sanit.*, **30**: 36—39
(in Russian).

ARHIPOVA, O. G., DJATLOVA, N. M., KABAČNIK, M. I., MEDVED', T. J.,
& MEDYNCEV, V. V. (1968) [On some regularities governing the formation of metal complexes by using polyamino-polyphosphonic acids and
elimination of metals from the organism.] *Bjull. eksp. Biol. Med.*, **12;**
66—69 (in Russian).

ASHIZAWA, R. (1927) [Report on an autopsy case of chronic manganese
poisoning.] *Jpn. J. med. Sci. Trans.*, **7** (1): 173—191 (in German).

BAADER, E. W. (1932) [Manganese poisoning in dry cell battery factories.]
Arch. Gewerbepathol. Gewerbehyg., **4**: 101—116 (in German).

BABSKIJ, E. B. & DONSKIH, E. H. (1972) [Nature of antagonistic effects
of manganese and nickel ions on action potentials of myocardial muscle
fibres.] *Dokl. Akad. Nauk.*, **27** (5): 1250—1258 (in Russian).

BARBEAU, A., INOUE, N., & CLOUTIER, T. (1976) Role of manganese
in dystonia. In: Eldridge, R. & Fahn, S., ed. *Advances in Neurology,*
New York, Raven Press, pp. 339—352.

BARBER, D. A. & LEE, R. B. (1974) The effect of micro-organisms on the
absorption of manganese by plants. *New Phytol.*, **73**: 97—106.

BARBOŘIK, M. & SEHNALOVÁ, H. (1967) [Blood levels and urinary excretion in electric arc welders. The effect of NaCa$_2$ EDTA on manganese
elimination in urine.] *Prac. Lék.*, **19** (4): 155—158 (in Czechk).

BARTELS, T. T. & WILSON, C. E. (1969) Determination of methylcyclopentadienyl manganese tricarbonyl in JP-4 fuel by atomic absorption
spectrophotometry. *At. Absorpt. Newsl.*, **8**: 3—5.

BAXTER, D. J., SMITH, W. O., & KLEIN, G. C. (1965) Some effects of
acute manganese excess in rats. *Proc. Soc. Exp. Biol. Med.*, **119**: 966—
970.

BAZELON, M., PAINE, R. S., COWIE, V. A., HUNT, P., HOUCK, J. C., &
MAHANAND, D. (1967) Reversal of hypotonia in infants with Down's
syndrome by administration of 5-hydroxytryptophan. *Lancet*, **1**: 1130—
1133.

BECKER, D. A. & MAIENTHAL, E. J. (1975) *Evaluation and research of
methodology for the national environmental specimen bank. Final
Report to US Environmental Protection Agency by US Department of
Commerce,* Washington, DC, National Bureau of Standards (Report
No. EPA/NBS/IGA-D5-0568).

BEGAK, O. Ju., KUKUŠKIN, Ju.N., NIKOLAEV, G. I., & POKROVSKAJA,
K. A. (1972) [The effect of acids and some elements on the determina-
tion of manganese by atomic absorption using an air-acetylene flame.]
Z. prikl. Himii, **45** (10): 2188—2191 (in Russian).

BEK, F., JANOVŠKOVA, J., & MOLDAN, B. (1972) [Direct determination
of manganese and strontium in the blood serum using a Perkin-Almer
HGA-70 graphite cuvette.] *Chem. Listy,* **66** (8): 867—75 (in Czech).

BELOBRAGINA, G. V. (1972) [The development of nodular structures in the
lungs in pneumoconiosis caused by condensed aerosols containing silicon
dioxide and manganese oxides.] In: *Kombinirovannoe dejstvie himičeskih in fizičeskih faktorov proizvodstvennoj sredy,* Sverdlovsk, pp.
128—138 (in Russian).

BELOBRAGINA, G. V. & EL'NICNYH, L. N. (1969) [Characteristics of the
changes occurring in parenchymatous organs in pneumoconiosis caused
by condensed aerosols containing silicon dioxide and manganese oxides.]

In: *Klinika, patogenez i profilaktika prfzabolevanii himičeskoj etiologii na predprijatijah cvetnoj i černoj metallurgii*, Sverdlovsk, Part II, pp. 107—114 (in Russian).

BELOBRAGINA, G. V., EL'NICNYH, L. N., & BABUSKINA, L. C. (1969) [On pneumosclerosis caused by condensed aerosols containing silicon dioxide and manganese oxides.] In: *Klinika, patogenez i profilaktika profzabolevanii himičeskoj etiologii na predprijatijah cvetnoj i černoj metallurgii*, Sverdlovsk, Part II, pp. 95—106 (in Russian).

BELZ, R. (1960) [The amounts of iron, copper, manganese and cobalt in average diets of various age groups in the Netherlands.] *Voeding*, **21**: 236—251 (in Dutch).

BENTLEY, P. J. (1967) Mechanism of action of neurohypophyseal hormones; actions of manganese and zinc on the permeability of the toad bladder. *J. Endocrinol.*, **39**: 493—506.

BERNHEIMER, H., BIRKMAYER, W., HORNYKIEWICZ, O., JELLINGER, K., & SEITELBERGER, F. (1973) Brain dopamine and the syndromes of Parkinson and Huntington. Clinical, morphological and neurochemical correlations. *J. neurol. Sci.*, **20**: 415—455.

BERTINCHAMPS, A. J., MILLER, S. T., & COTZIAS, G. C. (1966) Interdependence of routes excreting manganese. *Am. J. Physiol.*, **211**: 217—224.

BERTOGLIO — RIOLO, C., FULLE —SOLDI, T., & SPINI, G. (1972) [Determination of Cu (II), Cd (II), Pb (II), Zn (II), Cr (VI) in industrial waste waters by oscillopolarography.] *Ann. Chim.* **62**: 730—739 (in Italian).

BEŠČETNOVA, E. I., SAMOJLOV, V. V., KUPČIK, G. L., & SAMILKINA, N. C. (1968) [Concentrations of nickel, manganese, molybdenum, vanadium, titanium, copper and lead in the water of the lower reaches and delta of the Volga.] *Gig. i Sanit.*, **33** (8): 105 (in Russian).

BETHARD, W. F., OLEHY, D. A., & SCHMITT, R. A. (1964) The use of neutron activation analysis for the quantitation of selected cations in human blood. In: *L'analyse par radioactivation et ses applications aux sciences biologiques. III Colloque International de Biologie de Saclay, Septembre 1963*, Paris, Presses Universitaires de France, pp. 379—393.

BIRKS, L. S., GILFRICH, J. V., & BURKHALTER, P. G. (1972) *Development of X-ray fluorescence spectroscopy for elemental analysis of particulate matter in the atmosphere and in source emission, Washington, US* Environmental Protection agency, Office of Research and Monitoring, 43 pp. (EPA-R2-72-063).

BOCKMAN, P. W. K. (1939) [The Sauda pneumonia from the point of view of occupational hygiene.] *Nord. Med.*, **3**: 2548—2552 (in Norwegian).

van BOGAERT, L. & DALLEMAGNE, M. J. (1946) Approche expérimentale des troubles du manganisme. *Mschr. Psychiatr. Neurol.*, **111**: 60—89.

BONILLA, E. & DIEZ-EWALD, M. (1974) Effect of L-Dopa on brain concentration of dopamine and homovanillic acid in rats after chronic manganese chloride administration. *J. Neurochem.*, **22**: 297—299.

BOUQUIAUX, J. (1974) *Non-organic micropollutants of the environment. Vol. 2. Detailed list of levels present in the environment.* Luxembourg, Commission of the European Communities, pp. 194—221 (Document V/F/1966/74e).

BOWEN, H. J. M. (1956) The determination of manganese in biological material by activation analysis with a note on the gamma spectrum of blood. *J. nucl. Energy.* **3**: 18—24.

BRAR, S. S., NELSON, D. M., KANABROCKI, E. L., MOORE, C. E., BURNHAM, C. D., & HATTORI, D. M. (1970) Thermal neutron activation analysis of particulate matter in surface air of the Chicago metropolitan area. One-minute irradiations. *Environ. Sci. Technol.*, **4**: 50—54.

BRITTON, A. A. & COTZIAS, G. C. (1966) Dependence of manganese turnover on intake. *Am. J. Physiol.*, **211**: 203—206.

BROOKS, R. R. & RUMSBY, M. G. (1965) The biogeochemistry of trace element uptake by some New Zealand bivalves. *Limnol. Oceanogr.*, **10**: 521—527.

BROOKS, R. R. & RUMSEY, D. (1974) Heavy metals in some New Zealand commercial sea fishes. *N. Z. J. Mar. Freshwater Res.*, **8**: 155—166.

BROWN, E., SKOUGSTAD, M. W., & FISHMAN, M. J. (1970) Methods for collection and analysis of water samples for dissolved minerals and gases. Part. I. Sampling. In.: *Techniques of Water Resources Investigations of the United States Geological Survey*, Book 5, Washington, DC, US Government Printing Office, pp. 4—18.

BRYAN, D. E. (1970) *Development of nuclear analytical techniques for oil slick identification (Phase 1).* Work done under AEC contract No. AT (904—3)-167 by Gulf General Atomic (Report No. 9889).

BUGAEVA, M. A. (1969) [On the determination of lead and manganese in blood by spectrum analysis.] In: *Novoe v oblasti prom-sanhimii,* Moscow, Medicina, pp. 239—242 (in Russian).

BURNETT, W. T., Jr, BIGELOW, R. R., KIMBALL, A. W., & SHEPPARD, C. W. (1952) Radiomanganese studies on the mouse, rat and pancreatic fistula dog. *Am. J. Physiol.*, **168**: 620—625.

BUTT, E. M., NUSBAUM, R. E., GILMOUR, T. C., DIDIO, S. L., & SISTER MARIANO (1964) Trace metal levels in human serum and blood. *Arch. environ. Health,* **8**: 52—57.

BUTTIN, G. & KORNBERG, A. (1966) Enzymatic synthesis of deoxyribonucleic acid. XXI. Utilization of deoxyribonucleoside triphosphates by *Escherichia coli* cells. *J. biol. Chem.,* **241** (22): 5419—5427.

BYHOVSKAJA, M. S., GINZBURG, S. L., & HALIZOVA, O. D. (1966) [Cyclopentadienyltricarbonylmanganese.] In: *Metody opredelenija rednyh vescest v vozduhe,* Moscow, Goshimizdat, pp. 176—178 (in Russian).

CANAVAN, M. M. & DRINKER, C. K. (1934) Chronic manganese poisoning. Report of a case, with autopsy. *Arch. Neurol. Psychiatr.,* **32**: 501—513.

CARLBERG, J. R., CRABLE, J. V., LIMTIACA, L. P., NORRIS, H. B., HOLTZ, J. L., MAUER, P., & WOLOWICZ, F. R. (1971) Total dust, coal, free silica and trace metal concentrations in bituminous coal miners' lungs. *Am. Ind. Hyg. Assoc. J.,* **32**: 432—440.

CASAMAJOR, L. (1913) An unusual form of mineral poisoning affecting the nervous system; manganese. *J. Am. Med. Assoc.* **69**: 646—649.

CERETELI, M. N. & KIPIANI, S. L. (1971) Activity of the coagulative and anticoagulative system of the blood in Mn poisoning. In: *Documentation for the I All-Union Conference on Early Diagnosis and Prophylaxis of Occupational Diseases of Chemical Etiology,* Moscow, Ministry of Public Health, pp. 36—37.

CHANDRA, S. V. & SRIVASTAVA, S. P. (1970) Experimental production of early brain lesions by parenteral administration of manganese chloride. *Acta Pharmacol. Toxicol.,* **28**: 177—183.

CHANDRA, S. V., SETH, P. K., & MANKESHWAR, J. K. (1974) Manganese poisoning; clinical and biochemical observations. *Environ. Res.,* **7**: 374—380.

CHEESEMAN, R. V. & WILSON, A. L. (1972) *A method for the determination of manganese in water.* Marlow, England, Water Research Association (TP 85).

CHESTER, R. & HUGHES, M. J. (1967) A chemical technique for the separation of ferro-manganese minerals, carbonate minerals and absorbed trace elements from pelagic sediments. *Chem. Geol.,* **2**: 249—262.

CHOLAK, J. & HUBBARD, D. M. (1960) Determination of manganese in air and biological material. *Am. Ind. Hyg. Assoc. J.,* **21**: 356—360.

CIKRT, M. (1972) Biliary excretion of ^{203}Hg, ^{64}Cu ^{52}Mn and ^{210}Pb in the rat. *Br. J. ind. Med.,* **29**: 74—80.

CIKRT, M. (1973) Enterohepatic circulation of ^{64}Cu, ^{52}Mn and ^{203}Hg in rats. *Arch. Toxicol.,* **31**: 51—59.

CIKRT, M. & VOSTÁL, J. (1969) Study of manganese resorption *in vitro* through intestinal wall. *Int. Z klin. Pharmakol. Ther. Toxikol.*, **3**: 280—285.

COHN, M. & TOWNSEND, J. (1954) A study of manganous complexes by paramagnetic resonance absorption. *Nature (Lond)*, **173**: 1090—1091.

COMENS, P. (1956) Manganese depletion as an etiological factor in hydralazine disease. *Am. J. Med.*, **20**: 944—945.

CONSOLAZIO, C. F., NELSON, R. A., MATOUSH, L. O., HUGHES, R. C., & URONE, P. (1964) *The trace mineral losses in sweat*, Denver, CO, US Army Medical Research and Nutrition Laboratory, 14 pp. (Report No. 284).

COOK, D. G., FAHN, S., & BRAIT, K. A. (1974) Chronic manganese intoxication. *Arch. Neurol.*, **30**: 59—64.

COOPER, J. A. (1973) Comparison of particle and photon — excited X-ray fluorescence applied to trace element measurements of environmental samples. *Nucl. Instr. Met.*, **106**: 525—539.

COTZIAS, G. C. (1958) Manganese in health and disease. *Phys. Rev.*, **38**: 503—533.

COTZIAS, G. C. (1962) Manganese. In: Comar & Bronner, ed. *Mineral metabolism: An advanced treatise*, New York, Academic Press, Vol. 2, Part B, pp. 403—442.

COTZIAS, G. C. (1966) Manganese, melanins and the extrapyramidal system. *J. Neurosurg.*, **24**: 170—180.

COTZIAS, G. C. (1969) Metabolic modification of some neurological disorders. *J. Am. Med. Assoc.*, **210**: 1255—1262.

COTZIAS, G. C. & PAPAVASILIOU, P. S. (1962) State of binding of natural manganese in human cerebrospinal fluid, blood and plasma. *Nature (Lond.)* **195**: 823—834.

COTZIAS, G. C., PAPAVASILIOU, P. S., & MILLER, S. T. (1964) Manganese in melanins. *Nature (Lond.)*, **201**: 1228—1229.

COTZIAS, G. C., MILLER, S. T., & EDWARDS, J. (1966) Neutron activation analysis: the stability of manganese concentrations in human blood and serum. *J. lab. clin. Med.*, **67**: 836—849.

COTZIAS, G. C., HORIUCHI, K., FUENZALIDO, S., & MENA, I. (1968) Chronic manganese poisoning. Clearance of tissue manganese concentrations with persistence of the neurological picture. *Neurology*, **18**: 376—382.

COUGHANOWR, D. R. & KRAUSE, F. E. (1965) The reaction of SO_2 and O_2 in aqueous solutions of $MnSO_4$. *Ind. eng. Chem. Fundam.*, **4**: 61—66.

COULSTON, F. & GRIFFIN, T. (1976) *Inhalation toxicology of airborne particulate manganese in rhesus monkeys*, Institute of Comparative and Human toxicology, Albany Medical College, and New Mexico, International Center of Environmental Safety, Holloman Air Force Base, 97 pp. (Final report of US EPA Contract No. 68-02-0710).

CURRAN, G. L. (1954) Effect of certain transition group elements on hepatic synthesis of cholesterol in the rat. *J. biol. Chem.*, **210**: 765—770.

DAIRMAN, W. & UDENFRIEND, S. (1971) Decrease in adrenal tyrosine hydroxylase and increase in norepinephrine synthesis in rats given L-dopa. *Science*, **171**: 1022—1024.

DAMS, R., ROBBINS, J. A., RAHM, K. A., & WINCHESTER, J. W. (1970) Non-destructive neutron activation analysis of air pollution particulates. *Anal. Chem.*, **42**: 861—867.

DASTUR, D. K., MANGHANI, D. K., RAGHAVENDRAN, K. V., & JEE-JEEBHOY, K. N. (1969) Dstribution and fate of [54]Mn in rat with special reference to the CNS. *Q. J. exp. Physiol.*, **54** (3): 322—331.

DASTUR, D. K., MANGHANI, D. K., & RAGHAVENDRAN, K. V. (1971) Distribution and fate of [54]Mn in the monkey: studies of different parts of the central nervous system and other organs. *J. clin. Invest.*, **50**: 9—20.

DATE, S. (1960) The experimental studies on Mn intoxication. *J. Kumamoto Med. Soc.*, **34**: 159—182.

DAVIS, W. E. & ASSOCIATES (1971) *National inventory of sources and emissions: manganese-1968*. NC, US Environmental Protection Agency, Office of Air Programmes (Report No. APTD-1509).

DAVYDOVA, V. I. (1969) [The combined effect of manganese compounds and fluorine.] In: *Klinika, patogenez i profilaktika profzabolevanij himčeskoj etiologii na predprijatijah cvetnoj i černoj metallurgii*, Sverdlovsk, Part II, pp. 100—106 (in Russian).

DAVYDOVA, V. I., SABEL'NIKOVA, N. I., PANYČEVA, E. N., & MIHAJLOV, V. A. (1971) [On the toxicity of dust containing manganese and fluorine.] In: *Voprosy gigieny fisiologii truda i profpatologii*, Sverdlovsk, pp. 45—46, (in Russian).

DE, H. N. (1949) Copper and manganese metabolism with typical Indian diets and assessment of their requirement for Indian adult. *Ind. J. med. Res.*, **37**: 301—309.

DEMEREC, M., BERTANI, G., & FLINT, J. (1951) A survey of chemicals for mutagenic action on *E. coli. Am. Nat.*, **85**: 119—136.

DESMOND, E. A. (1975) *The environmental implications of manganese as an alternate antiknock. Discussion of SAE Paper No. 750926*, presented at the 1975 SAE Automobile Engineering Meeting, Detroit, MI, October 1975.

DOBRYNINA, O. Ju. & DAVIDJAN, L. G. (1969) [Changes with age in the levels of iron, manganese, copper and cobalt in the organs of healthy men.] *Med. Z. Usb.*, **12**: 47—50 (in Russian).

DOISY, E. A. (1973) Micronutrient controls on biosynthesis of clotting proteins and cholesterol. In: Hemphill, D.D., ed. *Trace substances in environmental health — VI*, Columbia, MO, University of Missouri, pp. 193—199.

DOKUČAEVA, V. F. & SKVORTSOVA, N. N. (1966) [On the problem of the biological effects of low concentrations of inhaled manganese.] In: *[Biological effects and hygienic significance of atmospheric concentrations.]* pp. 173—185 (in Russian).

DONALDSON, J., ST PIERRE, T., MINNICH J. L., & BARBEAU, A. (1973) Determination of Na^+, K^+, Mg^{2+}, Cu^{2+}, Zn^{2+} and Mn^2 in rat brain regions. *Can. J. Biochem.*, **51**: 87—92.

DONSKIH, E. A. & MUKUMOV, M. R. (1974) [The action of manganese ions on automatic activity of myocardial fibres induced by nickel ions.] *Bjull. Eksp. Biol. Med* **78** (9): 7—9 (in Russian).

DRESIA, H. & SPOHR, F. (1971) Experience with the radiometric dust-measuring unit "Beta Staubmeter". *Staub, Reinhalt. Luft*, **31** (6): 19—27.

DURFOR, C. N. & BECKER, E. (1964) Public water supplies of the 100 largest cities in the US, 1962. *I. Am. Water Works Assoc.*, **56** (3): 237—246.

DURHAM, N. N. & WYSS, O. (1957) Modified method of determining mutation rates in bacteria. *J. Bacteriol.*, **74**: 548—552.

DURUM, W. H. & HAFFTY, J. (1961) *Occurrence of minor elements in water*, Washington, DC, US Geological Survey, 11 pp. (Circular 445).

EL ALFY, S., LAQUA, R., & MASSMANN, H. (1973) [Spectrochemical analysis with a glow discharge lamp as a light source. III. Development and description of an universal method for the determination of the main constituents in electrically non-conducting powder samples.] *J. anal. Chem.*, **263**: 1—14 (in German).

ELDERFIELD, H. (1972) Compositional variations in the manganese oxide component of marine sediments. *Nature (Lond)*, **237** (70): 110—112.

ELISEEVA, T.N. (1973) [Micronutrient fertilizers in pickling liquors. The treatment of waste waters and industrial atmospheric emissions.] In: *Tezicy dokladov Volgogradskogo Medinskogo Instituta*, pp. 45—47 (in Russian).

ELLIS, G. H., SMITH, S. E., & GATES, E. M. (1947) Further studies of manganese deficiency in the rabbit. *J. Nutr.*, **34**: 21—31.

EL'NIČNYH, L. N. (1969) [Changes occurring in the synapses of the cerebral cortex during administration of dusts containing silicon dioxide and manganese oxides.] In: *Klinika, patogenez i profilaktika profzabolevanii himiceskoj etiologii na predprijatijah cvetnoj i černoj metallurgii*, Sverdlovsk, Part II, pp. 115—118 (in Russian).

ELSTAD, D. (1939a) [Factory smoke containing manganese as predisposing cause in epidemics of pneumonia in an industrial district.] *Nord. Med.*, **3:** 2527—2533, 2544—2548 (in Norwegian).

ELSTAD, D. (1939b) [Observations on manganese pneumonia.] In: *Proceedings of the VIII International Congress of Accident Medicine and Occupational Diseases, Frankfurt, AM, 26—30 September, 1938*, Vol.2, Leipzig, Georg Thieme Verlag, pp. 1014—1022 (in German).

EMARA, A. M., EL-GHAWABI, S. H., MADKOUR, O. I., & EL-SAMRA, G. H. (1971) Chronic manganese poisoning in the dry battery industry. *Br. J. ind. Med.*, **28:** 78—82.

ENGEL, R. W., PRICE, N. O., & MILLER, R. F. (1967) Copper, manganese, cobalt, and molybdenum balance in pre-adolescent girls. *J. Nutr.*, **92:** 197—204.

ENVIRONMENT AGENCY, JAPAN (1972) *[Notification of Director of Air Quality Bureau, June 1, 1972 Sampling and measurement method of suspended particulate matter.]* (in Japanese).

ENVIRONMENT AGENCY JAPAN (1975) [*Results of measurements at National Air Sampling Stations during 1973 (fiscal year).*] (in Japanese).

ERMAN, M. I. (1972) [Electric welding of low-carbon and low-alloy steel in shielding gases; hygienic and occupational-pathological aspects.] *Gig. Tr. prof. Zabol.*, **16** (5): 26—30 (in Russian).

ETHYL CORPORATION (1971) *The effect of manganese on the oxidation of sulfur dioxide*, Detroit, Ethyl Corp. 5 pp.

ETHYL CORPORATION (1974) *Methylcyclopentadienylmanganese tricarbonyl (MMT), an anti-knock agent for unleaded gasoline* (Status report, 2nd ed.), Ferndale, MI, Ethyl Corp. Research Lab. (Rep. ER-452).

von EULER, U.S. (1965) Catecholamines in nerve and organ granules. In: von Euler, U.S., Rosell, S., & Uvnäs, B., ed. *Mechanisms of release of biogenic amines*, London, Pergamon Press, pp. 211—222.

EVERSON, G. J. & SHRADER, R. E. (1968) Abnormal glucose tolerance in manganese deficient guinea pigs. *J. Nutr.*, **94:** 89—94.

FAGGAN, J. E., BAILIE, J. D., DESMOND, E. A., & LENANE, D. L. (1975) An evaluation of manganese as an antiknock in unleaded gasoline. Presented at the *1975 SAE Automobile Engineering Meeting, Detroit, MI October 13—17, 1975*.

FAULL, R. L. M. & LAVERTY, R. (1969) Changes in dopamine levels in the corpus striatum following lesions in the substantia nigra. *Exp. Neurol.*, **23:** 332—340.

FAURBYE, A. (1970) The structural and biochemical basis of movement disorders in treatment with neuroleptic drugs and in extrapyramidal diseases. *Comp. Psychiatr.*, **11** (3): 205—225.

FERNANDEZ, A. A., SOBEL, C., & JACOBS, S. L. (1963) Sensitive method for the determination of submicrogram quantities of manganese and its application to human serum. *Anal. Chem.*, **35:** 1721—1724.

FISHBEIN, L. (1973) Manganese, cobalt, nickel and cadmium. I. Manganese. In: *Chromatography of environmental hazards*, New York, American Elsevier Publishing Company, Vol. 2, pp. 37—42.

FLEMING, G. A. (1974) Trace elements in plants with particular reference to pasture species. *Outlook Agric.*, **4** (6): 270—285.

FLINN, R. H., NEAL, P. A., REINHART, W. H., DALLAVALLE, J. M., FULTON, W. B., & DOOLEY, A. E. (1940) *Chronic manganese poisoning in an ore-crushing mill*, Washington, DC, US Government Printing Office, pp. 1—77 (Public Health Bull. No. 247).

FORE, H. & MORTON, R. A. (1952) Microdetermination of manganese in biological materials by a modified catalytic method. *Biochem. J.*, **51**: 594—598.

GENTEL, J. E., MANARY, O. J., & VALENTA, J. C. (1974a) *Development of a methodology for the assessment of the effects of fuels and additives on the control devices*, Washington, DC, US Environmental Protection Agency (EPA-650/2-74-060).

GENTEL, J. E., MANARY, O. J., & VALENTA, J. C. (1974b) *Determination of effect of particulate exhaust emissions of additives and impurities in gasoline*, Washington, DC, US Environmental Protection Agency (EPA-650/2-74-061).

GEORGII, H.-W. & MÜLLER, J. (1974) [Heavy metal aerosols in metropolitan air.] *Schriftenr. Vereins Wasser, Boden Lufthyg.*, **42**: 39—50 (in German).

GEORGII, H.-W., JENDRICKE, U., JOST, D., & MÜLLER, J. (1974) The distribution of heavy metals in clean and polluted atmospheres. *Staub Reinhalt Luft*, **34**: 16—18.

GIAQUE, R. D., GOULDING, F. S., JAKLEVIC, J. M., & PEHL, R. H. (1973) Trace element determination with semiconductor detector X-ray spectrometers. *Anal. Chem.*, **45**: 671—681.

GOLDSTEIN, M., ANAGNOSTE, B., BATTISTA, A. F., OWEN, W. S., & NAKATANI, S. (1969) Studies of the amines in the striatum in monkeys with nigral lesions; the disposition, biosynthesis and metabolites of (H3) dopamine and (C14) serotonin in the striatum. *J. Neurochem.*, **16**: 645—653.

GOULDING, F. S. & JAKLEVIC, J. M. (1973) *X-ray fluorescence spectrometer for airborne particulate monitoring*. Washington, DC, US Environmental Protection Agency (EPA-R2-73-182).

GOULDING, F. S., WALTON, J. T., & PEHL, R. H. (1970) Recent results on the cytoelectronic feedback preamplifier. *IEEE Trans. Nucl. Sci.*, **NS-17**: 218—225.

GRAHAM, J. A., GARDNER, D. E., WALTERS, M. D., & COFFIN, D. L. (1975) Effect of trace metals on phagocytosis by alveolar macrophages. *Infect. Immunol.*, **11**: 1278—1283.

GREENBERG, D. M., COPP, D. H., & CUTHBERTSON, E. M. (1943) Studies in mineral metabolism with the aid of artificial radioactive isotopes. VII. The distribution and excretion, particularly by way of the bile of iron, cobalt and manganese. *J. biol. Chem.*, **147**: 749—756.

GUTHRIE, B. E. (1975) Chromium, manganese, copper, zinc and cadmium content of New Zealand foods. *N. Z. med. J.*, **82**: 418—424.

HAGENFELDT, K., PLANTIN, L. O., & DICZFALSY, E. (1973) Trace elements in the human endometrium. II. Zinc, copper and manganese levels in the endometrium, cervical mucus and plasma. *Acta Endocrinol.*, **72**: 115—126.

HAKIMOVA, A. M., ZELENKOVA, N. P., & PANČENKO, E. I. (1969) [Changes in the thyroid gland of rats with increasing amounts of manganese in the diet.] *Gig. i Sanit.*, **34** (1): 113—114 (in Russian).

HANCOCK, R. G. V., EVANS, D. J. R., & FRITZE, K. (1973) Manganese proteins in blood plasma. *Biochim. Biophys. Acta*, **320**: 486—493.

HANLON, D. P., GALE, T. F., & FERM, V. H. (1975) Permeability of the syrian hamster placenta to manganous ions during early embryogenesis. *J. Reprod. Fertil.*, **44**: 109—112.

HARMS, U. (1974) [Determination of the transition metals manganese, iron, cobalt, copper and zinc in river fishes with help of X-ray fluorescence analysis and flameless atomic absorption spetroscopy.] *Arch. Fischereiwiss.*, **25** (1/2): 63—74 (in German).

HARTMAN, R. H., MATRONE, G. & WISE, G. H. (1955) Effect of high dietary manganese on hemoglobin formation. *J. Nutr.*, **57**: 429—439.

HASEGAWA, N. & IJICHI, R. (1973) [Atomic absorption spectrophotometry

— problems in the preparation of the sample and its determinations.]
J. occup. Health Eng., **13**: 38—46 (in Japanese).

HATAMOV, S., BADALOV, N. Ju., & RUSTAMOV, R. (1972) [Determination of potassium, manganese, phosphorus, sodium and chlorine in the cotton plant by neutron activation analysis.] *Tr. Samarkandskogo Inst.*, **223**: 88—93 (in Russian).

HAŻARADZE, R. E. (1961) [Hygienic background for determining the maximum permissible concentration of manganese in water basins.] *Gig. i Sanit.*, **26** (12): 8—14 (in Russian).

HEGDE, B., GRIFFITH, G. C., & BUTT, E. M. (1961) Tissue and serum manganese levels in evaluation of heart muscle damage. A comparison with SGOT (26738). *Proc. Soc. Exp. Biol. Med.*, **107**: 734—737.

HILL, R. M. & HOLTKAMP, D. E. (1954) Storage of dietary manganese and thiamine in the rat. *J. Nutr.*, **53**: 73—82.

HILL, R. M., HOLTKAMP, D. E., BUCHANAN, A. R., & RUTLEDGE, E. K. (1950) Manganese deficiency in rats in relation to ataxia and loss of equilibrium. *J. Nutr.*, **41**: 359—371.

HORIGUCHI, S., UTSONOMIYA, T., KASAHARA, A., SHINAGAWA, K., IYODA, K., TANAKA, N., TSUYAMA, N., IKUTOMI, H., & MIURA, K. (1966) [A survey of the actual conditions of factories handling manganese compounds.] *Jpn. J. ind. Health*, **8** (6): 333—342 (in Japanese).

HORIUCHI, K., HORIGUCHI, S., TANAKA, N., & SHINAGAWA, K. (1967) Manganese contents in the whole blood, urine and feces of a healthy Japanese population. *Osaka City med. J.*, **13**: 151—163.

HORIUCHI, K., HORIGUCHI, S., SHINAGAWA, K., UTSUNOMIYA, T., & TSUYAMA, Y. (1970) On the significance of manganese contents in the whole blood and urine of manganese handlers. *Osaka City med. J.*, **16**: 29—37.

HUBUTIJA, V. A. (1972) [Distribution and retention of unipolar electrically charged industrial manganese dust in the respiratory organs.] *Gig. Tr. prof. Zabol.*, **2**: 27—31 (in Russian).

HUGHES, E. R. & COTZIAS, G. C. (1960) Manganese metabolism and adrenacortical activity; an isotropic study. *Fed. Proc.*, **19**: 249.

HUGHES, E. R., MILLER, S. T., & COTZIAS, G. C. (1966) Tissue concentrations of manganese and adrenal function. *Am. J. Physiol.*, **211**: 207—210.

HUNTINGDON RESEARCH CENTER (1975) *Evaluation of the chronic inhalation toxicity associated with a manganese aerosol produced from the combustion of methylcyclopentadienyl manganese tricarbonyl (MMT)*, 213 pp. (Final report submitted to Ethyl Corporation, Project no. 731—339).

HURLEY, L. S. (1968) Genetic-nutritional interactions concerning manganese. In: Hemphill, D.D., ed. *Trace substances in environmental health — II*, Columbia, MO, University of Missouri, pp. 41—51.

HURN, R. W., ALLSUP, J. R., & COX, F. (1974) *Effect of gasoline additives on gaseous emissions*, Washington, DC, US Environmental Protection Agency, 64 pp. (EPA-650/2-75-014).

HYSELL, D. K., MOORE, W. Jr, STARA, J. F., MILLER, R., & CAMPBELL, K. I. (1974) Oral toxicity of methylcyclopentadienyl manganese tricarbonyl (MMT) in rats. *Environ. Res.*, **7**: 158—168.

IARC WORKING GROUP (1976) *IARC Monographs on the evaluation of carcinogenic risk of chemicals to man.* Lyons, International Agency for Research on Cancer, Vol. 12, pp. 137—149.

IMAM, Z. & CHANDRA, S. V. (1975) Histochemical alterations in rabbit testis produced by manganese chloride. *Toxicol. appl. Pharmacol.*, **32**: 534—544.

INTERNATIONAL COMMISSION ON RADIOLOGICAL PROTECTION (1975) *Report of the Task Group on Reference Man*, Oxford, Pergamon Press, pp. 361—364, 393—394.

ITAKURA, T. & TAJIMA, T. (1972) [*A survey of environmental pollution by manganese dust from a ferro-manganese plant in Kanazawa.]* Ishikawa Prefecture, Institute for Hygiene and Environmental Pollution, Annual Report, Vol. 9, pp. 40—50 (in Japanese).

ITOYAMA, T. (1971) [Total iron and manganese in the piestic ground waters in the main artesian wells at Takamatsu city.] *J. Water Waste,* **152:** 5—11 (in Japanese).

JAPAN ENVIRONMENTAL SANITATION CENTRE (1974) [*Report on the survey of air pollution in special environments.*] (in Japanese).

JÄRVINEN, R. & AHLSTRÖM, A. (1975) Effect of the dietary manganese level on tissue manganese, iron, copper and zinc concentrations in female rats and their fetuses. *Med. Biol.,* **53:** 93—99.

JOHANSSON, T. B., VAN GRIEKEN, R. E., NELSON, J. W., & WINCHESTER, J. W. (1975) Elemental trace analysis of small samples by proton-induced X-ray emission. *Anal. Chem.,* **47:** 855—860.

JONDERKO, G. (1965) [Calcium, magnesium, inorganic phosphorus, sodium, potassium and iron levels in blood serum in the course of acute experimental manganese poisoning.] *Med. Pr.,* **16:** 288—292 (in Polish).

JONDERKO, G. & SZCZUREK, Z. (1967) [Pathomorphological studies of the internal organs in experimental manganese poisoning.] *Arch. Gewerbepathol. Gewerbehyg.,* **23:** 106—116 (in German).

JONDERKO, G., KUJAWSKA, A., & LANGAUER-LEWOWICKA, H. (1971) [Studies on the early symptoms of manganese toxicity.] *Med. Pr.,* **22** (1): 1—10 (in Polish).

JONDERKO, G., CZEKANSKA, D., TWARDOWSKI, Z., & TYRNA E. (1973a) [Effects of occupational exposure to manganese on the development of atherosclerosis.] *Med. Pr.,* **24** (6): 589—599 (in Polish).

JONDERKO, G., TWARDOWSKI, Z., & TYRNA, E. (1973b) [Effect of manganese exposure upon sickness absenteeism and morbidity.] *Med. Pr.* **24** (6): 629—638 (in Polish).

JONDERKO, G., KUJAWSKA, A., & LANGAUER-LEWOWICKA, H. (1974) [Effect of interruption of occupational contact with manganese upon neurological and biochemical symptoms of the toxic effect of manganese.] *Med. Pr.,* **25** (6): 543—548 (in Polish).

JONES, P. G. W., HENRY, J. L., & FOLKARD, A. R. (1973) *The distribution of selected trace metals in the water of the North Sea 1971—73,* Lowestoft, International Council for the Exploration of the Sea (ICES), Fisheries Improvement Committee, Fisheries Laboratory, 13 pp. (C.M. 1973/C: 5).

KAGAMIMORI, S., MAKINO, T., HIRAMARU, Y., KAWANO, S., KATO, T., NOGAWA, K., KOBAYASHI, E., SAKAMOTO, M., FUKUSHIMA, M., ISHIZAKI, A., KANAGAWA, K., & AZAMI, S. (1973) [Epidemiological studies on disturbance of respiratory system caused by manganese air pollution (Report 2). Improvement of respiratory functions after setting up dust collectors.] *Jpn. J. Public Health,* **20** (8): 413—421 (in Japanese).

KANABROCKI, E. L., FIELDS, T., DECKER, C. F., CASE, L. F., MILLER, E. B., KAPLAN, F., & OESTER, Y. T. (1964) Neutron activation studies of biological fluids; manganese and copper. *Int. J. appl. Radiat. Isot.,* **15:** 175—190.

KAPLAN, R. W. (1962) [Problems in testing pharmaceutical products, additives and other chemicals for their mutagenic action.] *Naturwissenschaften,* **49:** 457—462 (in German).

KAWAMURA, R., IKUTA, H., FUKUZUMI, S., YAMADA, R., & TSUBAKI, S. (1941) Intoxication by manganese in well water. *Kitasato Arch. exp. Med.,* **18:** 145—169.

KEANE, J. R. & FISHER, E. M. R. (1968) Analysis of trace elements in airborne particulates by neutron activation and X-ray spectrometry. *Atmos. Environ.,* **2:** 603—614.

KEHOE, R. A., CHOLAK, J., & STORY, R. V. (1940) A spectrochemical study of the normal ranges of concentration of certain trace metals in biological materials. *J. Nutr.,* **19:** 579—592.

KESIĆ, B. & HÄUSLER, V. (1954) Heamatological investigation on workers exposed to manganese dust. *Am. Med. Assoc. Arch. ind. Health occup. Med.*, **10**: 336—343.

KIMURA, Y., MINURA, S., & HIGUCHI, I. (1969) [Nature of underground water in Japan.] *Ann. Rep. Tokyo Metrop. Res. Lab., Public Health*, **20**: 119—121 (in Japanese).

KITAMURA, S., SUMINO, K., HAYAKAWA, K., & SHIBATA, T. (1974) [Heavy metals in normal Japanese subjects (Amounts of 15 heavy metals in 30 subjects).] *Environ. Health Rep.*, **28**: 29—60 (in Japanese).

KLAASSEN, C. D. (1974) Biliary excretion of manganese in rats, rabbits and dogs. *Toxicol. appl. Pharmacol.*, **29**: 458—467.

KLAWANS, H., Jr, ILAHI, M. M., & SHENKER, D. (1970) Theoretical implications of the use of L-dopa in Parkinsonism. *Acta neurol. Scand.*, **46**: 409—441.

KLEINKOPF, M. D. (1960) Spectrographic determination of trace elements in lake waters of Northern Maine. *Geol. Soc. Am. Bull.*, **71**: 1231—1242.

KOBERT, R. (1883) [On the pharmacology of manganese and iron.] *Arch. exp. Pathol. Pharmakol.*, **16**: 361—392 (in German).

KOCMOND, W. C., YANG, J. Y., & DAVIS, J. A. (1975) *A methodology for determining the effects of fuel and additive combustion products on atmospheric visibility,* Buffalo, New York, Calspan Corporation (Calspan Rep. No. NA. 5300-M-1).

KOLESNIKOVA, V. G., KASATKINA, L. A., MALAHOVA, O. P., MUSYČENKO, V. I., & BABOV, D. M. (1973) [Concentrations of the trace elements copper and manganese in river water, soils and plants in the southern Ukraine.] *Gig. i Sanit.*, **38** (3): 110—111 (in Russian).

KOLOMIJEEVA, M. G. (1970) *[Trace elements in medicine.]* Moscow, Medicina, 287 pp. (in Russian).

KOSAI, M. F. & BOYLE, A. J. (1956) Ethylenediaminetetraacetic acid in manganese poisoning of rats. A preliminary study. *Ind. Med. Surg.*, **25**: 1—3.

KOSHIDA, Y., KATO, M., & HARA, T. (1963) Autoradiographic observations of manganese in adult and embryo mice. *Q. J. exp. Physiol.*, **48**: 370—378.

KOSHIDA, Y., KATO, M., & HARA, T. (1965) Distribution of radiomanganese in embryonic tissues of the mouse. *Annot. zool. Jpn*, **38** (1): 1—7.

KOSTERLITZ, H. W. & WATERFIELD, A. A. (1972) Effects of calcium and manganese on acetylcholine release from the myenteric plexus of guinea pig and rabbit ileum. *Br. J. Pharmacol.*, **45**: 157—158.

KOSTIAL, K., LANDEKA, M., & ŠLAT, B. (1974) Manganese ions and synoptic transmission in the superior cervical ganglion of the rat. *Br. J. Pharmacol.*, **51**: 231—235.

KOZUKA, T., MIYAGI, K., SATO, T., MORI, R., & SHIKANAI, H. (1971) [Cd and Mn in streams of the upper Iwaki river.] *Bull. Fac. Educ. Hirosaki Univ.*, **24B**: 7—13 (in Japanese).

KRONER, R. C. & KOPP, J. F. (1965) Trace elements in six water systems of the United States. *Am. Water Works Assoc. J.*, **57**: 150—156.

LAGERWERFF, J. V. (1967) Heavy-metal contamination of soils. In: Brady, N.C., ed. *Agriculture and the quality of our environment*, Norwood, MA, Plimpton Press, pp. 343—364.

LANDIS, D. A., GOULDING, F. S., PEHL, R. H., & WALTON, J. T. (1971) Pulsed feedback techniques for semiconductor detector radiation spectrometers. *IEEE Trans. Nucl. Sci., N. S.* **18**: 115—124.

LASSITER, J. W., MORTON, J. D., & MILLER, W. J. (1970) Influence of manganese on skeletal development in the sheep and rat. In: Mills, C.F., ed. *Trace element metabolism in animals, Proceedings of WAAP/IBP International Symposium, Aberdeen, July 1969*, Edinburgh, E & S Livingstone, pp. 130—132.

LAZRUS, A. L., LORANGE, E., & LODGE, J. P. Jr (1970) Lead and other metal ions in United States precipitation. *Environ. Sci. Technol.*, **4**: 55—58.

LEE, R. E., GORANSON, S. S., ENRIONE, R. E., & MORGAN, G. B. (1972) NASN cascade impactor network. Part II: Size distribution of trace metal components. *Environ. Sci. Technol.*, **6**: 1025—1030.

LERNER, M. O., GONČAROV, V. V., HESINA, A. Y., & BAKALEINIK, A. M. (1974) [Effects of the nature of fuel and the additive cyclopentadie-nyltricarbonyl-manganese on the concentration of benzo pyrene in exhaust gases of a carburator motor.] *Ekspl.-Tekh. Svoistva Primen. Avtomov. Topl. Smaz. Mater. Spetszhidk.*, No 8, pp. 51—54 (in Russian) (Abstract in *Air Pollut. ind. Hyg.*, 1974, **8**: 365).

LETAVET, A. A. (1973) [Manganese.] In: *Professional'nye bolezni*, Moscow (in Russian).

LEVANOVSKAJA, A. I. & NEIZVESTNOVA, E. M. (1972) [Morphological and histochemical characteristics of the adrenal glands on exposure to manganese chloride and generalized vibration.] In: *Kombinirovannoe dejstvie himićeskih i fizičeskih faktorov proizvodstvennonj sredy*, Sverdlovsk, pp. 84—90 (in Russian).

LEVINA, E. N. & ROBAČEVSKAJA, E. G. (1955) [Changes in the lung tissues under intratracheal injections of manganese oxides.] *Gig. i Sanit.* **20** (1): 25—28 (in Russian).

LEVINA, E. N. & TČEKUNOVA, M. P. (1969) [Manganese on monoamin-oxidase activity.] *Vopr. Med. Him.*, **15**: 634—637 (in Russian).

LEVITT, M., SPECTOR, S., SJOERDSMA, A., & UDENFRIEND, S. (1965) Elucidation of the rate-limiting step in norepinephrine biosynthesis in the perfused guinea pig heart *J. Pharmacol. exp. Ther.*, **148**: 1—8.

LINDBERG, O. & ERNSTER, L. (1954) Manganese a co-factor of oxidative phosphorylation. *Nature (Lond.)* **173**: 1037—1038.

LLOYD DAVIES, T. A. (1946) Manganese pneumonitis. *Br. J. ind. Med.*, **3**: 111—135.

LLOYD DAVIES, T. A. & HARDING, H. E. (1949) Manganese pneumonitis. Further clinical and experimental observations. *Br. J. ind. Med.*, **6**: 82—90.

L'VOV, B. V. (1961) The analytical use of atomic absorption spectra. *Spectrochim. Acta* **17**: 761—770.

LYONS, M. & INSKO, W. M. (1937) Chondrodystrophy in the chick embryo produced by manganese deficiency in the diet of the hen. *Kentucky Agric. Exp. Stn Bull.*, **371**: 63—75.

MAHONEY, J. P. & SMALL, W. J. (1968) Studies on manganese. III. The biological half-life of radiomangenese in man and factors which affect this half-life. *J. clin. Invest.*, **47**: 643—653.

MAHONEY, J. P., SARGENT, K., GRELAND, M., & SMALL, W. (1969) Studies on manganese. I. Determination in serum by atomic absorption spectrophotometry. *Clin. Chem.*, **15**: 312—322.

MAIGRETTER, R. Z., EHRLICH, R., FENTERS, J. D., & GARDNER, D. E. (1976) Potentiating effects of manganese dioxide on experimental respiratory infections. *Environ. Res.*, **11**: 386—391.

MANDELL, A. J. & SPOONER, C. E. (1968) Psycho-chemical research studies in man. *Science*, **162**: 1442—1453.

MANDŽGALADZE, R. N. (1966a) [On the effect of manganese compounds on sexual functions in the male rat.] *Vopr. Gig. Tr. Profpatol.*, **10**: 191—195 (in Russian).

MANDŽGALADZE, R. N. (1966b) [The effect of manganese compounds on the estrous cycle and embryogenesis in experimental animals.] *Vopr. Gig. Tr. Profpatol.*, **10**: 219—225 (in Russian).

MANDZGALADZE, R. N. (196c) [On the mutagenic properties of manganese compounds.] *Vopr. Gig. Tr. Profpatol.* **10**: 225—226 (in Russian).

MANDŽGALADZE, R. N. (1967) [Some clinical and experimental data on the effect of manganese compounds on sexual function.] *Vopr. Gig. Tr. Profpatol.* **11:** 126—130 (in Russian).

MANDŽGALADZE, R. N. & VASAKIDZE, M. I. (1966) [The effect of small doses of manganese compounds, nitrogenous organomercury pesticides and some anticoagulants on chromosome rearrangements in white rat bone marrow cells.] *Vopr. Gig. Tr. Profpatol.* **10:** 209—212 (in Russian).

MATRONE, G., HARTMAN, R. H., & CLAWSON, A. J. (1959) Studies of a manganese-iron antagonism in the nutrition of rabbits and baby pigs. *J. Nutr.,* **67:** 309—317.

MATTESON M. J., STÖBER, W., & LUTHER, H. (1969) Kinetics of the oxidation of sulfur dioxide by aerosols of manganese sulfate. *Ind. eng. Chem. Fundam.,* **8:** 677—687.

MAVRINSKAJA, L. F., MIHAJLOV, V. A., & DAVYDOVA, V. I. (1972) [Morphological changes in the nerve components of somatic muscle on combined exposure to manganese and carbon monoxide.] In: *Kombinirovannoe dejstvie himiceskih i fiziceskih faktorov proizvodstvennoj sredy.* Sverdlovsk, pp. 109—114 (in Russian).

MAYNARD, L. S. & COTZIAS, G. C. (1955) The partition of manganese among organs and intracellular organelles of the rat. *J. biol. Chem.,* **214:** 489—495.

MAYNARD, L. S. & FINK, S. (1956) The influence of chelation on radio-manganese excretion in man and mouse. *J. clin. Invest.,* **35:** 831—836.

McKAY, H. A. C. (1971) The atmospheric oxidation of sulfur dioxide in water droplets in the presence of ammonia. *Atmos. Environ.,* **5:** 7—14.

McLEOD, B. E. & ROBINSON, M. F. (1972a) Dietary intake of manganese by New Zealand infants during the first six months of life. *Br. J. Nutr.,* **27:** 229—232.

McLEOD, B. E. & ROBINSON, M. F. (1972b) Metabolic balance of manganese in young women. *Br. J. Nutr.,* **27:** 221—227.

MELLA, J. (1924) The experimental production of basal ganglion symptomatology in *Macacus rhesus. Arch. Neurol. Psychiatr.,* **11:** 405—417.

MENA, I. (1974) The role of manganese in human disease. *Ann. clin. lab. Sci.,* **4** (6): 487—491.

MENA, I. & COTZIAS, G. C. (1970) *Manganese poisoning: a metabolic disorder.* Final report on US PHS research project. pp. 21—26 (Grant No. EC-213-07).

MENA, I., MARIN, O., FUENZALIDA, S., COTZIAS, G. C. (1967) Chronic manganese poisoning. Clinical picture and manganese turnover. *Neurology,* **17:** 128—136.

MENA, I., HORIUCHI, K., BURKE, K., & COTZIAS, G. C. (1969) Chronic manganese poisoning. Individual susceptibility and absorption of iron. *Neurology,* **19:** 1000—1006.

MENA, I., COURT, J., FUENZALIDA, S., PAPAVASILIOU, P. S., & COTZIAS, G. C. (1970) Modification of chronic manganese poisoning: treatment with *L*-dopa or 5-OH tryptophane. *New Engl. J. Med.,* **282:** 5—10.

MENA, I., HORIUCHI, K., & LOPEZ, G. (1974) Factors enhancing entrance of manganese into the brain: Iron deficiency and age. *J. nucl. Med.,* **15:** 516.

MÉRANGER, J. C. & SMITH, D. C. (1972) The heavy metal content of a typical Canadian diet. *Can. J. public Health,* **63:** 53—57.

MIHAJLOV, V. A. (1969) [Manganese poisoning in steel casting shop workers.] In: *Klinika, patogenez i profilaktika profzabolevanij himičeskoj etiologii na predprijatijah cvetnoj i černoj metallurgii.* Sverdlovsk, pp. 190—206 (in Russian).

MIHAJLOV, V. A. (1971) [Some urgent problems of interest in the pathogenesis and therapy of manganesotoxicosis.] *Gig. Tr. prof. Zabol.,* **6:** 14—19 (in Russian).

MIHAJLOV, V. A., KLJAČINA, K. N., BELJAEVA, L. N., SADILOVA, M.

S., KAZANCEVA, T. I., TROP, F. C., KLEINER, A.-M., BELOBRAGINA, G. B., OKONISNIKOVA, I. E., DORINOVSKAJA, A. P., NEIZVESTNOVA, E. M., SIGOVA, N. V., & BLOHIN, V. (1967) [Research results and future lines of investigation with regard to pathogenesis, therapy and individual preventive care in the case of poisoning by compounds of manganese, chromium and fluoride.] In: *Obščie voprosy promyšlennoj toksikologii*, Moscow, Institute of Industrial Hygiene and Occupational Diseases, pp. 124—127 (in Russian).

MIHAJLOV, V. A., DAVYDOVA, V. I., & EL'NICNYH, L. I. (1969) [Experimental data on aspects of the combined effect of manganese compounds and carbon monoxide.] In: *Klinika, patogenez i profilaktika profzabolevanij himičeskoj etiologii na predprijatijah cvetnoj i černoj metallurgii, Sverdlovsk, Part II*, pp. 104—108 (in Russian).

MILLER, R. S., MILDVAN, A. S., CHANG, H. C., EASTERDAY, R. L., MARUYAMA, H., & LANE, M. D. (1968) The enzymatic carboxylation of phosphoenolpyruvate. IV. The binding of manganese and substrates by phosphoenolpyruvate carboxy-kinase and phosphoenol pyruvate carboxylase. *J. biol. Chem.*, **243**: 6030—6040.

MILLER, S. T., COTZIAS, G. C., & EVERT, H. A. (1975) Control of tissue manganese; initial absence and sudden emergence of excretion in the neonatal mouse. *Am. J. Physiol.*, **229**: 1080—1084.

MINERALS YEARBOOK, (1977) Washington, DC, US Department of Interior, Bureau of Mines.

MINISTRY OF INTERNATIONAL TRADE AND INDUSTRY AND MINISTRY OF HEALTH AND WELFARE, JAPAN (1969) *[Report on the survey of environmental manganese concentrations in Yokkaichi district.]* Tokyo, 139 pp. (in Japanese).

MITCHELL, H. H. & HAMILTON, T. S. (1949) The dermal excretion under controlled environmental conditions of nitrogen and minerals in human subjects, with particular reference to calcium and iron. *J. biol. Chem.*, **178**: 345—361.

MITCHELL, R. L. (1964) Trace elements in soils. In: Bear, F. E., ed. *Chemistry of the soil*, 2nd ed., New York, Reinhold Publ. Corp., pp. 320—321.

MITCHELL, R. L. (1971) *Trace elements in soils, Aberdeen, Macaulay* Institute for Soil Research, pp. 8—20 (Tech. Bull. No. 20).

MJAČINA, L.Ja. (1972) [Monoaminoxidase activity in nerve components and muscle tissue in manganese poisoning.] In: *Kombinirovannoe dejstvie himičeskij i fizičeskij faktorov proizvodstvennoj sredy:* Sverdlovsk, pp. 114—116 (in Russian).

MOORE, W. Jr., HALL, L., CROCKER, W., ADAMS, J., & STARA, J. F., (1974) Metabolic aspects of methylcyclopentadienyl manganese tricarbonyl in rats. *Environ. Res.*, **8**: 171—177.

MOORE, W., HYSELL, D., MILLER, R., MALANCHUK, M., HINNERS, R., YANG, Y., & STARA, J. F. (1975) Exposure of laboratory animals to atmospheric manganese from automotive emissions. *Environ. Res.*, **9**: 274—284.

MORAN, J. B. (1975) The environmental implications of manganese as an alternate antiknock. *Paper, presented at the 1975 SAE Automobile Engineering Meeting, Detroit, MI, October, 1975*, (No. SAE 750926).

MORAN, J. B., BALDWIN, M. J., MANARY, O. J., & VALENTA, J. C. (1972) *Effect of fuel additives on the chemical and physical characteristics of particulate emissions in automotive exhaust, Washington, DC, US* Environmental Protection Agency, 345 pp. (EPA-R2-72-066).

MORGEN, E. A., VLASOV, N. A., & KOZEMJAKINA, L. A. (1972) [A kinetic method for determining trace amounts of manganese according to the decreasing fluorescence of the morin complex of beryllium.]   Z. *Anal. Him.*, **27** (10): 2064—2067 (in Russian).

MORRIS, A. W. (1974) Seasonal variation of dissolved metals on inshore waters of the Menai straits. *Mar. Pollut. Bull.*, **5** (4): 54—59.

MOSENDZ, S. A. (1968) [Effect of electric-welding dust on the metabolism of nitrogenous substances in the tissues.] *Vrach. Delo.*, **5**: 109—111 (in Russian).

MOURI, T. (1973) [Experimental study of inhalation of manganese dust.] *Shikoku Acta Med.*, **29** (2): 118—129 (in Japanese).

MUHTAROVA, M., BOBEV, G., COHEN, E., & GENOVA, S. (1969) [A rapid drop wise quantitative method for determination of manganese aerosols in the air of the working environment.] *Works Inst. Labour Prot. occup. Dis.*, **18** (11): 257—261 (in Bulgarian).

MURADOV, V. G. & MURADOVA, O. N. (1972) [Determination of the absolute concentration of manganese atoms in the gas phase by an atomic absorption method.] *Z. Prikl. Spektrosk.*, **17** (6): 1102—1104 (in Russian).

MUSAEVA, L. D. & KOZLOVA, G. I. (1973) [The accumulation of manganese in bilberries as a result of the environmental condition of the leaves.] *Nauc. dokl. vyssh. sk. biol. n.*, **7**: 71—76 (in Russian).

MUSTAFA, S. J. & CHANDRA, S. V. (1971) Levels of 5-hydroxytryptamine, dopamine, and norepinephrine in whole brain of rabbits in chronic manganese toxicity. *J. Neurochem.*, **18**: 931—933.

NAKAGAWA, T. (1968) [A study on the manganese content in daily food consumption in Japan.] *J. Osaka City Cent.*, **17** (9—10): 401—424 (in Japanese).

NAKAMURA, Y. & OSADA, M. (1957) [Studies on trace elements in food and beverages. III. Manganese content in green tea.] *Ann. Rep. Kyoritsu Coll. Pharm.*, **3**: 19—21 (in Japanese).

NAS/NRC (1973) *Manganese*, Washington, DC, National Academy of Sciences, National Research Council, pp. 1—191.

NEFF, N. H., BARRET, R. E., & COSTA, E. (1969) Selective depletion of caudate nucleus dopamine and serotonin during chronic manganese dioxide administration to squirrel monkeys. *Experientia (Basel)*, **25**: 1140—1141.

NEIZVESTNOVA, E. M. (1972a) [An investigation into the effect on the body of manganese dioxide and generalized vibration.] In: *Kombinirovannoe dejstvie himičeskih i fizičeskih faktorov proizvodstvennoj sredy*, Sverdlovsk, pp. 76—80 (in Russian).

NEIZVESTNOVA, E. M. (1972b) [The toxicity of manganous chloride under conditions of generalized vibration.] In: *Kombinirovannoe dejstvie himičeskih i fizičeskih faktorov proizvodstvennoj sredy*, Sverdlovsk, pp. 84—90 (in Russian).

NICHOL, I., HORSNAIL, R. F., & WEBB, J. S. (1967) Geochemical patterns in stream sediment related to precipitation of manganese oxides. *Trans Inst. Min. Metall. Sect. B*, **76**: 113—115.

NISHIYAMA, K., SUZUKI, Y., FUJII, N., YANO, H., MIYAI, T., & ONISHI, K. (1975) [Effect of long-term exposure to manganese particles. Part 2. Periodical observation of respiratory organs in monkeys and mice.] *Jpn. J. Hyg.*, **30**: 117 (in Japanese).

NOGAWA, K., KOBAYASHI, E., SAKAMOTO, M., FUKUSHIMA, M., ISHIZAKI, A., KAGAMIMORI, S., MAKINO, T., HIRAMARU, Y., KAWANO, S., KATO, T., KANAGAWA, K., & AZAMI, S. (1973) [Epidemiological studies on disturbance of respiratory system caused by manganese air pollution. Report 1: Effects on respiratory system of junior high school students.] *Jpn. J. public Health*, **20** (6): 315—326. (in Japanese).

NORDBERG, G. F., ed. (1976) *Effects and dose-response relationships of toxic metals*, Amsterdam, Elsevier, 559 pp.

NORTH, B. B., LEICHSENRING, J. M., & NORRIS, L. M. (1960) Manganese metabolism in college women. *J. Nutr.*, **72:** 217—223.

OBOLENSKAJA, L. I., KAMENSKAJA, L. S., & BUZJUKINA, V. V. (1971) [An atomic absorption method for determining zinc, manganese, iron and magnesium in plants and soil.] *Agrohimija*, **1:** 136—142 (in Russian).

OLEHY, D. A., SCHMITT, R. A., & BETHARD, W. F. (1966) Neutron activation analysis of magnesium, calcium, strontium, barium, manganese, cobalt, copper, zinc, sodium, and potassium in human erythrocytes and plasma. *J. nucl. Med.*, **7:** 917—927.

OSIPOVA, I. A., D'JAKOVA, I. N., & URMANCEEVA, T. G. (1968) [Manganese parkinsonism in experiments on macaque and rhesus monkeys (Part I).] *Gig. Tr. prof. Zabol.*, **12** (8): 48—49 (in Russian).

PANIĆ, B. (1967) [*In vitro* binding of manganese to serum transferrin in cattle.] *Acta Vet. Sand.*, **8:** 228—233.

PAPAVASILIOU, P. S. & COTZIAS, G. C. (1961) Neutron activation analysis; the determination of manganese. *J. biol. Chem.*, **236:** 2365—2369.

PAPAVASILIOU, P. S., MILLER, S. T., & COTZIAS, G. C. (1966) Role of liver in regulating distribution and excretion of manganese. *Am. J. Physiol.*, **211:** 211—216.

PARNITZKE, K. H. & PEIFFER, J. (1954) [On the clinical and pathological anatomy in chronic manganese poisoning.] *Arch. Psychiatr. Z. Neurol.*, **192:** 405—429 (in German).

PEÑALVER, R. (1955) Manganese poisoning: The 1954 Ramazzini oration. *Ind. Med. Surg.*, **24:** 1—7.

PENTSCHEW, A., EBNER, F. F., & KOVATCH, R. M. (1963) Experimental manganese encephalopathy in monkeys. A preliminary report. *J. Neuropathol. exp. Neurol.*, **22:** 488—499.

PEPIN, D., GARDES, A., PETIT, J., BERGER, J.-A., & GAILLARD, G. (1973) Étude spectrographique des éléments-traces des eaux minérales. I. Analyse directe du résidu sec. *Analysis*, **25:** 337—342.

PEREGUD, E. A. & GERNET, E. V. (1970) [Determination of manganese oxides and salts.] In: *Himiceskih analiz vozduha promyslennyh predprijatij*, Leningrad, Himija, pp. 342—343 (in Russian).

PFITZER, E. A., WITHERUP, S. O., LARSEN, E. E., & STEMMER, K. L. (1972) *Toxicity of methylcyclopentadienyl manganese tricarbonyl (MMT).* (Internal report to US Environmental Protection Agency. Kettering Laboratory, University of Cincinnati).

PLUMLEE, M. P., THRASHER, D. M., BEESON, W. M., ANDREWS, F. N., & PARKER, H. E. (1956) The effects of a manganese deficiency upon the growth, development and reproduction of swine. *J. anim. Sci.*, **15:** 354—367.

POČAŠEV, E. N. & NEIZVESTNOVA, E. M. (1972) [Morphological changes in some internal organs of white rats on exposure to a combination of manganese and vibration.] In: *Kombinirovannoe dejstvie himičeskih i fizičeskih faktorov proizvodstvennoj sredy*, Sverdlovsk, pp. 90—94 (in Russian).

POIRIER, L. J. & SOURKES, T. L. (1964) Influence of substantia nigra on the catecholamine content of the striatum. *Brain*, **88:** 181—194.

POIRIER, L. J., SINGH, P., BOUCHER, R., BOUVIER, G., OLIVIER, A., & LAROCHELLE, P. (1967a) Effect of brain lesion on striatal monamines in the cat. *Arch. Neurol. (Chic.)*, **17:** 601—608.

POIRIER, L. J., SINGH, P., SOURKES T. L., & BOUCHER, R. (1967b) Effect of amine precursors on the concentration of striatal dopamine and serotonin in cats with and without unilateral brainstem lesions. *Brain Res.*, **6:** 654—666.

POIRIER, L. J., McGEER, E. G., LAROCHELLE, L., McGEER, P. L. BEDARD, P., & BOUCHER, R. (1969) The effect of brain stem lesions on tyrosine and tryptophan hydroxylases in various structures of the telencephalon of the cat. *Brain. Res.*, **14**: 147—155.

POLLACK, S., GEORGE, J. N., REBA, R. C., KAUFMAN, R. M., & CROSBY, W. H. (1965) The absorption of nonferrous metals in iron deficiency. *J. clin. Invest.*, **44**: 1470—1473.

POVOLERI, F. (1974) [Bronchopneumonia and the production of ferro-manganese.] *Med. Lav.*, **38**: 30—34 (in Italian).

PRESTON, A., JEFFERIES, D. F., DUTTON, J. W. R., HARVEY, B. R., & STEELE, A. K. (1972) British isles coastal waters. The concentrations of selected heavy metals in sea water, suspended matter and biological indicators — a pilot survey. *Environ. Pollut.*, **3**: 69—82.

PRICE, N. O., BUNCE, G. E., & ENGEL, R. W. (1970) Copper, manganese, and zinc balance in pre-adolescent girls. *Am. J. clin. Nutr.*, **23**: 258—260.

REITH, J. W. S. (1970) Soil factors influencing the trace element content of herbage. In: Mills, C. F., ed. *Trace element metabolism in animals*, London, Livingstone, E. & S., pp. 410—412.

RHODES, J. R., PRADZYNSKI, A. H., HUNTER, C. B., PAYNE J. S., & LINDGREN, J. L. (1972) Energy dispersive X-ray fluorescence analysis of air particulates in Texas. *Environ. Sci. Technol.*, **6**: 922—927.

RIDDERVOLD, J. & HALVORSEN, K. (1943) Bacteriological investigations on pneumonia and pneumococcus carriers in Sauda, an isolated industrial community. *Acta Pathol. Microbiol. Scand.*, **20**: 272—298.

RILEY, J. P. & TAYLOR, D. (1968) Chelating resins for the concentration of trace elements from sea water and their analytical use in conjunction with atomic absorption spectrophotometry. *Anal. Chim. Acta*, **40**: 479—485.

RODIER, J. (1955) Manganese poisoning in Moroccan mines. *Br. J. ind. Med.*, **12**: 21—35.

ROŠČIN, A. V. (1971) [On some mechanisms of the toxic effects of metals.] In: *Material Vsesojuznoj konferencii po rannej diagnostike i profilaktike professional'nyh zabolevanij himiceskoj etiologii*, Moscow, Institute of Hygiene and Occupational Diseases, pp. 42—45 (in Russian).

ROSENSTOCK, H. A., SIMONS, D. G., & MEYER, J. S. (1971) Chronic manganism. Neurological and laboratory studies during treatment with levodopa. *J. Am. Med. Assoc.*, **217**: 1354—1359.

RUCH, R. R., GLUSKOTER, H. J., & SHIMP, N. F. (1973) *Occurrence and distribution of potentially volatile trace elements in coal:* An interim report, Illinois State Geological Survey, Environmental Geology Notes, No. 61, pp. 1—43.

RYLANDER, R. & BERGSTRÖM R. (1973) Particles and SO_2 — Synergistic effects for pulmonary damage. In: *Proceedings of the 3rd International Clean Air Congress, Düsseldorf, 1973*, Düsseldorf, VDI Verlag, pp. A23—A25.

RYLANDER, R., ÖHRSTRÖM, M., HELLSTRÖM, P. A., & BERGSTRÖM, R. (1971) SO_2 and particles — synergistic effects on guinea-pig lungs. In: Walton, W. H., ed. *Inhaled Particles III Working, Surrey, Unwin Brothers Ltd.*, pp. 535—571.

ŠABADAS, E. (1969) *[On the neurohumoral regulation of the daily cycle of the manganese level of the blood.]* In: *II Vsesojuznyh biohimiceskij S'ezd g. Taskent 1969. Tez. seke. soobč.* Tashkent, Izdatel'stvo FAN Uzbekskoj, SSR, Vol. 13, pp. 159—160 (in Russian).

SABNIS, C. V., YEANNAWAR, P. K., PAMPATTIWAR, V. L., & DESHPANDE, J. M. (1966) An environmental study of a ferro-manganese alloy concern. *Indian J. ind. Med.*, **(11—12)**: 207—222.

SAMOHVALOV, S. G., ČEBOTAREVA, N. A., & ZUKOVA, L. F. (1971) [Determination of available manganese in soils using formaldoxine.] *Him. Sel. Hoz.*, **9** (4) 63—65 (in Russian).

ŠARIĆ, M. (1972) Manganese exposure and respiratory impairment In: *XVII*

International Congress on Occupational Health, Buenos Aires, 17—23 September, 1972, pp. 101 (Abstract D2—16).

SARIĆ, M. (1978) Biological effects of manganese. Research Triangle Pask, NC, USA, US Environmental Protection Agency 152 pp. (Environmental Health Effects Research Series, EPA-600/1-78-001).

ŠARIĆ, M. & HRUSTIĆ, O. (1975) Exposure to air-borne manganese and arterial blood pressure. Environ. Res., 10: 314—318.

ŠARIĆ, M. & LUČIĆ-PALAIĆ, S. (1977) Possible synergism of exposure to airborne manganese and smoking habit in occurrence of respiratory symptoms. In: Inhaled Particles IV, Oxford & New York, Pergamon Press pp. 773—779.

SARIĆ, M., LUČIĆ-PALAIĆ, S., PAUKOVIĆ, R., & HOLETIĆ, A. (1974) [Respiratory effects of manganese.] Arh. hig. Rada, 25: 15—26 (in Croat).

SARIĆ, M., OFNER, E., & HOLETIĆ, A. (1975) Acute respiratory diseases in a manganese contaminated area. In: International Conference on Heavy Metals in the Environment Proceedings, Toronto, Canada, Institute foɪ Environmental Studies, University of Toronto, pp. 389—398.

SARIĆ, M., MARKIĆEVIĆ, A., & HRUSTIĆ, O. (1977) Occupational exposure to manganese. Br. J. ind. Med., 34: 114—118.

SCHAFER, D. F., STEPHENSON, D. V., BARAK, A. J., & SORRELL, M. F. (1974) Effects of ethanol on the transport of manganese by small intestine of the rat. J. Nutr., 104: 101—104.

SCHLAGE, C. & WORTBERG, B. (1972) Manganese in the diet of healthy preschool and school children. Acta Paediatr. Scand., 61: 648—652.

SCHROEDER, H. A. (1970) Manganese — Air Quality Monograph, Washington, DC, American Petroleum Institute, 34 pp. (No. 70—17).

SCHROEDER, H. A., PERRY, H. M., DENNIS, E. G., & MAHONEY, L. E. (1955) Pressor substances in arterial hypertension. V. Chemical and pharmacological characteristics of pherentasin. J. exp. Med., 102: 319—333.

SCHROEDER, H. A., BALASSA, D. D., & TIPTON, I. H. (1966) Essential trace elements in man; manganese, a study on homeostasis. J. chron. Dis., 19: 545—571.

SCHULER, P., OYANGUREN, H., MATURANA, V., VALENZUELA, A., CRUZ, E., PLAZA, V., SCHMIDT, E., & HADDAD, R. (1957) Manganese poisoning. Environmental and medical study at a Chilean mine. Ind. Med. Surg., 26: 167—173.

SHARMAN, D. F., POIRIER, L. J., MURPHY, G. F., & SOURKES, T. L. (1967) Homovanillic acid and dihydroxyphenylacetic acid in the striatum of monkeys with brain lesions. Can. J. Physiol. Pharmacol., 45: 57—62.

SHIMIZU, N. & MORIKAWA, N. (1957) Histochemical studies of succinic dehydrogenase of brain of mice, rats, guinea pigs and rabbits. J. Histochem. Cytochem., 5: 33—345.

SHRADER, R. E. & EVERSON, G. J. (1968) Pancreatic pathology in manganese-deficient guinea pigs. J. Nutr., 94: 269—281.

ŠIGAN, S. A. & VITVICKAJA, B. R. (1971) [Experimental substantiation of permissible residual concentrations of potassium permanganate in drinking water.] Gig. i Sanit., 36 (9): 15—18 (in Russian).

SIMMON, V. F. & LIGON, S. (1977) In vitro microbiological mutagenicity studies of ethyl corporation compounds. Interim report, California, Stanford Research Institute, 19 pp.

SLAVIN, W. (1968) Atomic absorption spectroscopy, New York, Interscience. Publishers Inc., pp. 129—130.

SLAVNOV, B. I. & MANDADZIEV, I. H. (1968) [A research into the effect of lead, manganese and zinc upon some indices of the reactivity of the body and on non-specific morbidity.] Gig. Tr. prof. Zabol., 12 (10): 22—26 (in Russian).

SMYTH, L. T., RUHF, R. C., WHITMAN, N. E., & DUGAN, T. (1973)

Clinical manganism and exposure to manganese in the production and processing of ferromanganese alloy. *J. occup. Med.*, **15**: 101—109.

SOMAN, S. D., PANDAY, V. K., JOSEPH, K. T., & RAUT, S. J. (1969) Daily intake of some major and trace elements. *Health Phys.*, **17**: 35—40.

SOURKES, T. L. & POIRIER, L. J. (1966) Neurochemical bases of tremor and other disorders of movement. *Can. Med. Assoc. J.*, **94**: 53—60.

SPENCER, D. W. & SACHS, P. L. (1970) Some aspects of the distribution, chemistry and mineralogy of suspended matter in the Gulf of Maine. *Mar. Geol.*, **9**: 117—136.

STADLER, H. (1936) [Histopathology of the brain resulting from manganese poisoning.] *Z. Ges. Neurol. Psychiatr.*, **154**: 62—76 (in German).

STOKINGER, H. E. (1962) The metals (excluding lead): Manganese, Mn. In: Patty, F. A., Fassett, D. W., & Irish, D. D., ed. *Industrial Hygiene and Toxicology*, 2nd revised edition, New York, Interscience Publishers, Inc., Vol. 2, pp. 1079—1089.

STONER, G. D., SHIMKIN, M. B., TROXELL, M. C., THOMPSON, T. L. & TERRY, L. S. (1976) Test for carcinogenicity of metallic compounds by the pulmonary tumor response in strain A mice. *Cancer Res.*, **36**: 1744—1747.

SULLIVAN, R. J. (1969) *Preliminary air pollution survey of manganese and its compounds. A literature review*, Raleigh, NC, National Air Pollution Control Administration, US Department of Health, Education and Welfare, 54 pp. (Publ. No. APTD 69—39).

SUNDERMAN, F. W. Jr, LAU, T. J., & CRALLEY, L. J. (1974) Inhibitory effect of manganese upon muscle tumorigenesis by nickel subsulfide. *Cancer Res.*, **34**: 92—95.

SUTHERLAND, E. W., ROBISON, G. A., & BUTCHER, R. W. (1968) Some aspects of the biological role of adenosine 3,5-monophosphate (cyclic AMP). *Circulation*, **37**: 279—306.

SUZUKI, Y. (1968) [Quantitative determination of manganese in plant leaves by atomic absorption spectrophotometry.] *Shikoku Acta Med.*, **24** (4): 453—456 (in Japanese).

SUZUKI, Y. (1970) [Environmental contamination by manganese.] *Jpn. J.ind. Health*, **12** (11): 529—533 (in Japanese).

SUZUKI, Y. (1974 [Studies on excessive oral intake of manganese. Part 2. Minimum dose for manganese accumulation in mouse organs.] *Shikoku Acta Med.*, **30**: 32—45 (in Japanese).

SUZUKI, Y., KAZIMOTO, M., KONDO, T., MATSUOKA, C., TANIOKA, Y., NISHIYAMA, S., HIROSE, A., KAWAKITA, H., & SHIMADA, K. (1968) [Determination of manganese in biological materials by atomic absorption spectrophotometry.] *Shikoku Acta Med.*, **24**: 449—456.

SUZUKI, Y., NISHIYAMA, K., SUZUKI, Y., & KURAMOTO, M. (1972) [Acute toxicity of dust from a new dry process of SO$_2$ removal in mice.] *Shikoku Acta Med.*, **28**: 339—343 (in Japanese).

SUZUKI, Y., NISHIYAMA, K., SUZUKI Y., KAJIMOTO, M., FUJII, N., NISHIDONO, N., KONDO, T., SOKABE, E., MATSUHASHI, M., UENO, E., & YAMADA, Y. (1973a) [The effects of chronic manganese exposure on ferromanganese workers (Part I).] *Shikoku Acta Med.*, **29**: 412—424 (in Japanese).

SUZUKI, Y., NISHIYAMA, K., SUZUKI, Y., KAJIMOTO, M., FUJII, N., NISHIDONO, N., KONDO, T., SOKABE, E., MATSUHASHI, M., UENO, E., & YAMADA, Y (1973b) [The effects of chronic manganese exposure on ferromanganese workers (Part 2).] *Shikoku Acta Med.*, **29**: 433—438 In Japanese).

SUZUKI, Y., NISHIYAMA, K., SUZUKI, Y., KAJIMOTO, M., FUJII, N., NISHIDONO, N., KONDO, T., SOKABE, E., MATSUHASHI, M., UENO, E., & YAMADA, Y. (1973c) [Blood and urinary manganese in ferromanganese workers.] *Shikoku Acta Med.*, **29**: 425—432 (in Japanese).

SUZUKI, Y., MOURI, T., SUZUKI, Y., NISHIYAMA, K., FUJII, N., &

YANO, H. (1975) [Study of subacute toxicity of manganese dioxide in monkeys.] *Tokushima J. exp. Med.*, **22**: 5—10 (in Japanese).

SWAINE, D. J. (1962) *The trace-element content of fertilizers.* Commonwealth Bureau of Soil Science, pp. 150—177 (Technical Communication No. 52).

TANAKA, S. & LIEBEN, J. (1969) Manganese poisoning and exposure in Pennsylvania. *Arch. environ. Health,* **19**: 674—684.

TASK GROUP ON METAL ACCUMULATION (1973) Accumulation of toxic metals with special reference to their absorption, excretion and biological half-lives. *Environ. Physiol. Biochem.,* **3**: 65—107.

TEPPER, L. B. (1961) Hazards to health: manganese. *New Engl. J. Med.,* **264**: 347—348.

TER HAAR, G. L., GRIFFING, M. E., BRANDT, M., OBERDING, D. G., & KAPRON, M. (1975) Methylcyclopentadienyl manganese tricarbonyl as an antiknock: Composition and fate of manganese exhaust products. *J. Air Pollut. Control Assoc.,* **25** (8): 858—860.

THIERS, R. E. & VALLEE, B. L. (1957) Distribution of metals in subcellular fractions of rat liver. *J. biol. Chem.,* **226**: 911—920.

THOMPSON, R. J., MORGAN, G. B., & PURDUE, L. J. (1970) Analysis of selected elements in atmospheric particulate matter by atomic absorption. *At. Absorpt. Newsl.,* **9**: 53—57.

THOMSON, A. B. R., OLATUNBOSYN, D. & VALBERG, L. S. (1971) Interrelation of intestinal transport system for manganese and iron. *J. lab. clin. Med.,* **78** (4): 642—655.

TICHÝ, M. & CIKRT, M. (1972) Manganese transfer into the bile in rats. *Arch. Toxicol.,* **29**: 51—58.

TICHÝ, M., HAVRDOVA, J., & CIKRT, M. (1971) [Determination of manganese in urine by atomic absorption spectrometry.] *Pr. Lek.,* **23**: 336—341 (in Czech).

TICHÝ, M., CIKRT, M., & HAVRDOVA, J. (1973) Manganese binding in rat bile. *Arch. Toxicol.,* **30**: 227—236.

TIPTON, I. H. & COOK, M. J. (1963) Trace elements in human tissue. Part II. Adult subjects from the United States. *Health Phys.,* **9**: 103—145.

TIPTON, I. H., COOK, M. J., STEINER, R. L., BOYE, C. A., PERRY, H. M. Jr, & SCHROEDER, H. A. (1976) Trace elements in human tissue. I. Methods. *Health Phys.,* **9**: 89—101.

TIPTON, I. H., STEWART, P. L., & DICKSON, J. (1969) Patterns of elemental excretion in long term balance studies. *Health Phys.,* **16**: 455—462.

TOPPING, G. (1969) Concentrations of Mn, Co, Cu, Fe, and Zn in the Northern Indian Ocean and Arabian Sea. *J. Mar. Res.,* **27**: 318—326.

TUREKIAN, N. K. (1969) The oceans, streams, and atmosphere. In: Wedephol, K .H., ed. *Handbook of geochemistry, Vol. 1,* New York, Springer-Verlag, pp. 297—323.

TUREKIAN, N. K. & WEDEPOHL, K. H. (1961) Distribution of the elements in some major units of the earth's crust. *Geol. Soc. Am. Bull.,* **72**: 175—191.

UENO, K. & OHARA, S. (1958) [On manganese poisoning in a steel smelting factory and loadtest of manganese.] *J. Labour Hyg. Iron. Steel Ind.,* **7** (4): 3—19 (in Japanese).

UNDERWOOD, E. J. (1971) Manganese. In: *Trace Elements in Human and Animal Nutrition,* 3rd ed., New York, Academic Press, pp. 177—207.

US ENVIRONMENTAL PROTECTION AGENCY (1971) Reference method for the determination of suspended particulates in the atmosphere. *Fed. Reg.,* **36** (84): Part II, 8191—8194.

US ENVIRONMENTAL PROTECTION AGENCY (1972a) *Atmosphere and in source emissions,* Washington, DC, Office of Research and Monitoring, 43 pp. (Report No. EPA-R-2-72-063).

US ENVIRONMENTAL PROTECTION AGENCY (1972b) *Air quality data for 1968 from the National Air Surveillance networks and contributing*

State and Local networks, Washington, US EPA, 231 pp. (EPA Report No. APTD-0978).

US ENVIRONMENTAL PROTECTION AGENCY (1973) *Air quality data for metals 1968 from the National Air Surveillance Networks*, Washington, US EPA, pp. 1-1-14-B (EPA Report No. APTD-1467).

US ENVIRONMENTAL PROTECTION AGENCY (1974) *Methods for chemical analyses of water and wastes*, Washington, US EPA, 298 pp. (EPA Report. No. EPA-600/6-72-002).

US ENVIRONMENTAL PROTECTION AGENCY (1975) *Scientific and technical assessment report on manganese*, Washington, US EPA, Office of Research and Development, 48 pp. (EPA Report No. EPA-600/6-75-002).

USOVIČ, A. T. (1967) [The application of polarography to the determination of trace elements in biological material.] In: *Mikroelementy v biosfere i ih primenenie v sel'skom hozjajstve i medicine Sibiri i Dal'nego Vostoka*, Ulan Ude, pp. 608—611 (in Russian).

VAN KOETSVELD, E. E. (1958) The manganese and copper contents of hair as an indication of the feeding condition of cattle regarding manganese and copper. *Tijdschr. Diergeneeskd.,* **83**: 229—236.

VAN ORMER, D. G. & PURDY, W. C. (1973) The determination of manganese in urine by atomic absorption spectrometry. *Anal. Chim. Acta,* **64**: 93—105.

VASILEVSKAJA, V. D. & BOGATYREV, A. G. (1970) [Trace elements in the soils of Western Tamir.] *Vestn. Mosk. Univ., ser. 6, Biol., poovovedenie,* **4**: 53—59 (in Russian).

VERSIECK, J., BARBIER, F., & HOSTE, J. (1975) Influence of myocardial infarction on serum manganese, copper, and zinc concentrations. *Clin. Chem.,* **21** (4): 578—581.

VERSIECK, J., SPEECKE, A., HOSTE, J., & BARBIER, F. (1973a) Trace contamination in biopsies of the liver. *Clin. Chem.,* **19**: 472—475.

VERSIECK, J., SPEECKE, A., HOSTE, H., & BARBIER, F. (1973b) Determination of manganese, copper and zinc in serum and packed blood sells by neutron activation analysis. *Z. klin. Chem. klin. Biochem.,* **11** (5): 193—196.

VERSIECK, J., BARBIER, F., SPEECKE, A., & HOSTE, J. (1974a) Normal manganese concentrations in human serum. *Acta Endocrinol.,* **76**: 783.

VERSIECK, J., BARBIER, F., SPEECKE, A., & HOSTE, J. (1974b) Manganese, copper and zinc concentrations in serum and packed blood cells during acute hepatitis, chronic hepatitis and posthepatic cirrhosis. *Clin. Chem.,* **20** (9): 1141—1145.

VOSS, H. (1939) [Progressive bulbar paralysis and amyotrophic lateral sclerosis from chronic manganese poisoning.] *Arch. Gewerbepathol. Gewerbehyg.,* **9**: 464—476 (in German).

VOSS, H. (1941) [Spinal cord and peripheral nervous system in chronic manganese poisoning.] *Arch. Gewerbepathol. Gewerbehyg.,* **10**: 550—568 (in German).

WALTERS, M. D., GARDNER, D. E., ARANYI, C., & COFFIN, D. L. (1975) Metal toxicity for rabbit alveolar macrophages *in vitro. Environ. Res.,* **9**: 32—47.

WASSERMANN, D. & WASSERMANN, M. (1977) The ultrastructure of the liver cell in subacute manganese administration. *Environ. Res.,* **14**: 379—390.

WATANABE, H., BERMAN, S., & RUSSEL, D. S. (1972) Determination of trace metals in water using X-ray fluorescence spectrometry. *Talanta,* **19**: (12) 1363—1375.

WEAST, R. C., ed. (1974) *Handbook of chemistry and physics,* 55th ed., Cleveland, CRC Press, pp. B107—B109.

WHITLOCK, C. M. Jr, AMUSO, S. J., & BITTENBENDER, J. B. (1966) Chronic neurological disease in two manganese steel workers. *Am. Ind. Hyg. Assoc. J.* **27**: 454—459.

WORLD HEALTH ORGANIZATION (1969) WHO Technical Report Series No. 417 (Pesticide Residues in food. Report of the 1968 Joint FAO/WHO Meeting), Geneva, WHO, 40 pp.

WORLD HEALTH ORGANIZATION (1973) WHO Technical Report Series, No. 532 (Trace elements in human nutrition — Report of a WHO Expert Committee), pp. 34—36.

WORLD HEALTH ORGANIZATION WORKING GROUP (1973) Technical document on manganese. In: *The hazards to health of persistent substances in water.* Annexes to the Report of a WHO Working Group, Coppenhaguen, WHO Regional Office for Europe, pp. 111—134.

WIDDOWSON, E. M. (1969) Trace elements in human development. In: Barltrop & Burland, ed. *Mineral metabolism in paediatrics,* Oxford, Blackwell, pp. 85—98.

WIDDOWSON, E. M., CHAN, H., HARRISON, G. E. & MILNER, R. G. D. (1972) Accumulation of Cu, Zn, Mn, Cr, and Co in the human liver before birth. *Biol. Neonate,* **20** (5—6): 360—367.

WILGUS, H. S. & PATTON, A. R. (1939) Factors affecting manganese utilization in the chicken. *J. Nutr.,* **18**: 35—45.

WITZLEBEN, C. L. (1969) Manganese-induced cholestasis: concurrent observations on bile flow rate and hepatic ultrastructure. *Am. J. Pathol.,* **57**: 617—625.

WYNTER, J. E. (1962) The prevention of manganese poisoning. *Ind. Med. Surg.,* **31**: 308—310.

YAMAMOTO, H. & SUZUKI, T. (1969) [Chemical structure of manganese compounds and their biological effects.] In: *[Proceedings of the 42nd Annual Meeting of the Japan Association of Industrial Health, 28—31 March, 1959.]* Fukuoka City, Japan, Japan Association of Industrial Health (in Japanese).

ZAJIC, J. E. (1969) *Microbial biochemistry,* New York, Academic Press, pp. 156—168.

ZAPADNJUK, V. I. & ZAIKA, M. U. (1970) [The effect of repeaated courses of administration of biotic doses of copper and manganese both singly and together, on some biochemical indices in mice.] *Farmakol. Toksikol.,* **5**: 606—607 (in Russian).

ŽERNAKOVA, T. V. (1967) [Correlation between iron, manganese and copper content in the blood serum of healthy individuals.] *Bjull. eksp. Biol. Med.,* **63**: 47—48 (in Russian).

ZOOK, G. E., GREENE, F. E., & MORRIS, E. R. (1970) Nutrient composition of selected wheats and wheat products. VI. Distribution of manganese, copper nickel, zinc, magnesium, lead, tin, cadmium, chromium and selenium as determined by atomic absorption spectroscopy and colorimetry. *Cereal Chem.,* **47**: 720—731.

ZUCKERMAN, J. J., ed. (1976) *Organotin compounds: New chemistry and applications,* Washington, DC, American Chemical Society.

ZUI, L. B. (1970) [Effect of pear varieties on the corrosion of cans made of electrolytically tin-plated sheet iron.] *Nauch. Tr. Vissh. Inst Khrant. Vkusova Prom. Plovdic.,* **17**: 312—320 (in Bulgarian).

WORLD HEALTH ORGANIZATION

CORRIGENDUM

ENVIRONMENTAL HEALTH CRITERIA
No. 13

Carbon Monoxide

Page 117, line 26:

Should read: adequate, though at higher CO-levels it overestimates [HbCO]. It has been used, for example, by Peterson & Stewart (1970)

Page 117:

Equation (6) should read: $[HbCO]_t = \dfrac{1}{A} \{e^{-kt}(A[HbCO]_o - V_{co}B - [CO]) + \dot{V}_{co}B + [CO]\}$

Page 117, last line:

Should read: t and $t = 0$, $[CO] = pI_{co}$, $A = \dfrac{\overline{p}CO_2}{M[HbO_2]}$, $B = (\dfrac{1}{D_L} + \dfrac{(pB - 47)}{\dot{V}_A})$ and $k = \dfrac{A}{\dot{V}_B . B}$